J. Lequeux, T. Encrenaz and F. Casoli

The Exoplanet Revolution

CURRENT NATURAL SCIENCES

EDP Sciences

Cover image: a planet just formed in the disk of gas and dust surrounding the dwarf star PDS 70. In this near-infrared, high-resolution image, the planet is seen as a bright point inside the remains of the disk. The star itself, at the center, is occulted by a mask. © European Southern Observatory, A. Müller et al.

Printed in France

© **2020, EDP Sciences**, 17 avenue du Hoggar, BP 112, Parc d'activités de Courtabœuf, 91944 Les Ulis Cedex A

This work is subject to copyright. All rights are reserved, whether the whole or part of the material is concerned, specifically the rights of translation, reprinting, re-use of illustrations, recitation, broad-casting, reproduction on microfilms or in other ways, and storage in data bank. Duplication of this publication or parts thereof is only permitted under the provisions of the French Copyright law of March 11, 1957. Violations fall under the prosecution act of the French Copyright law.

ISBN (print): 978-2-7598-2210-2 - ISBN (ebook): 978-2-7598-2211-9

Table of Contents

Introduction	1
Chapter 1. The prehistory of exoplanets	5
First ideas and speculations.	5
The evolution of concepts on the formation of the Solar System	8
The discovery of protoplanetary disks.	11
The first attempts to detect exoplanets.	12
Bibliography	17
Chapter 2. The first detections	19
An unexpected discovery: planets around a pulsar!.	19
1995: The first planet around a star like ours!.	21
The success of velocimetry.	27
Bibliography	31
Chapter 3. The method of transits	33
What is a planetary transit?.	33
Observations from the Earth	36
The Space Age.	37
The CoRoT mission	37
The Kepler Mission.	38
Primary and Secondary Transits.	40
Transmission spectroscopy (primary transit)	41
Emission spectroscopy (secondary transit).	43
Gravitational transits	44
Bibliography	46
Chapter 4. Detecting and viewing exoplanets	47
Advantages and limitations of indirect methods of detection	47
The transit method.	47
Exoplanets detected as gravitational microlenses	48
Exoplanets detected by velocimetry or astrometry	49
Direct observation: a very difficult problem.	50
Coronography.	51
The black-fringe interferometer	52
How to obtain perfect images: adaptive optics	53
Combined coronography and adaptive optics	55
A new track for the future: the search for exoplanets in the radio domain	57
Bibliography	57

Chapter 5. The variety of exoplanets — 59
  The outstanding results of the last twenty years. — 59
    A multitude of exoplanets — 59
    Giant exoplanets very close to their stars — 59
    Orbits of all kinds. — 60
    Many multiple systems — 61
    Planets around double stars. — 62
  The different classes of exoplanets — 62
    Hot and Cold Jupiters. — 64
    Super-Earths and Neptunes — 67
    Earths and Habitable Planets — 70
  Bibliography — 73

Chapter 6. The birth of stars and protoplanetary disks — 75
  Protostars, jets and disks. — 75
  The protoplanetary disks. — 78
  The ice lines in protoplanetary disks. — 82
  Planet-disk interactions. — 84
  Bibliography — 87

Chapter 7. Formation and evolution of planetary systems — 89
  The formation of planets. — 89
  The evolution of planetary systems: what does the Solar System teach us?. — 92
  Why no super-Earths in the Solar System?. — 98
  Expelled planets, isolated exoplanets — 98
  What future for the Solar System?. — 100
  What consequences for our understanding of exoplanetary systems?. — 102
  Bibliography — 104

Chapter 8. The physical nature of exoplanets — 107
  The observables. — 107
  The first measurements of the atmospheric composition of hot Jupiters. — 111
  Possible causes of departure from thermochemical equilibrium. — 114
  Clouds and mists on exoplanets — 116
  Spectroscopic measurements of super-Earths and exo-Neptunes in transit — 116
  Spectroscopy of exoplanets from the ground — 118
  Phase curves and atmospheric circulation of exoplanets — 120
  Bibliography — 121

Chapter 9. Around Exoplanets — 123
  The exocomets. — 124
  Giant rings around an exoplanet?. — 127
  A satellite? — 129
  Bibliography — 130

Chapter 10. Life on exoplanets? — 131
  What is life?. — 132
  The emergence of life on Earth. — 133
  Life elsewhere in the Solar System? — 141
  How to detect life on exoplanets?. — 144
  Bibliography — 149

Table of Contents V

Chapter 11. Exploring Exoplanets: What Prospects? **151**
    Habitable planets, inhabited planets: a difficult problem............ 151
    More planets, and smaller ones, to explore the diversity
        of planetary systems.................................. 154
    New light on giant planets: astrometry and imagery............... 159
    Studying the atmospheres of exoplanets 162
    Towards analogs of the Earth-Sun system: after 2030? 165
    Alpha Centauri: other planets?............................... 167

Chapter 12. Communicating with other worlds? **169**
    The probability of existence of advanced civilizations on other planets.. 169
    The SETI Projects 171
    Messages to the Universe................................... 174
    Bibliography 177

Appendix 1: The planets of the Solar System **179**

Appendix 2: A Selection of Exoplanets **181**

Appendix 3: Some useful data **189**

Appendix 4: The nomenclature of stars and exoplanets **191**

Glossary **193**

Acronyms **199**

Index **203**

Introduction

Among the great astronomical discoveries of the last century, the most exciting in the eyes of the general public was undoubtedly that of a planet around another star than the Sun. It dates back to 1995. To say the truth, it was expected that such a discovery would occur some day, because there was no reason why the Solar System should be the only one of its kind. From the time of Antiquity, then with Nicolas de Cues and Giordano Bruno at the end of the sixteenth century, there was speculation about the possibility of planetary systems similar to ours around other stars. It was only necessary that the techniques of observation had progressed sufficiently to allow the discovery. This is what differentiates the discovery of the extrasolar planets, as well as that of the Higgs boson in 2012 and the gravitational waves in 2015, from that of the unexpected astronomical objects that are the dark matter (discovered in 1937), the X-ray sources (1965), the quasars (1963), the fossil radiation from the Universe (1965), the pulsars (1965) and the sources of gamma-ray bursts (1968), to name only the most spectacular. The surprise was quite different: the new planet, which has a mass close to that of Jupiter, gravitates around the star 51 Pegasi in only four days and is twenty times less distant from this star than the Earth is from the Sun, which makes it a very hot object. There is nothing similar in the Solar System, which was immediately degraded from its status as a prototype of possible planetary systems. Soon other exoplanets (a neologism that seems accepted today) were discovered. They led to other surprises: some have very elongated elliptical orbits, which is not the case of any of the planets that are familiar to us. Super-Earths, and mini-Neptunes, with a mass intermediate between that of the Earth and that of Jupiter, are frequent, though unparalleled in the Solar System. We have also found planets orbiting around systems of double stars.

More than 4200 exoplanets have been discovered today. They show an astonishing variety in their masses, in their orbits and in their physical properties. Up to seven planets have been detected in some planetary systems. Most of the known exoplanets are much closer to their stars than the Earth is to the Sun, which was a big surprise; however, this may be only a consequence of the limits of our observations, for these planets are easier to detect than the distant ones. The latter, however, exist: they are the only ones whose image can be made without being too impeded by the light of the star, which can be a billion times more intense. There are even isolated planets circulating in the Milky Way: have they been ejected from a planetary system by some mechanism?

All this questioned the ideas we had about the birth and evolution of planetary systems. It was long believed with Laplace that the planets were at the distance of the star to which they were born, and formed an almost immutable system. Yet some researchers have shown in the 1980s that this could not be so because of the gravitational interaction between the planets and what remains of the gas and dust disk where they were born at the same time as the central star. Moreover, the discoverers of the first exoplanet immediately realized that it must have been formed far from the star, and then had initiated a migration that had moved it very close to this star. This observation, together with other unexpected properties of exoplanets, led to an intense theoretical activity, beginning with the properties of the Solar System that were thought to be well understood. The results are astonishing: it appears that the present aspect of the Solar System depends entirely on the behavior of its two biggest planets, Jupiter and Saturn, and would have been very different if only one had been present. Moreover, the Solar System, like other planetary systems, is not immutable, and could still undergo major transformations on the scale of several billion years: one could, for example, witness the ejection of a planet or a collision between two planets, an otherwise unpredictable behavior because it is dominated by chaos. Understanding the origin and structure of exoplanet systems, for which we have much less data than for the Solar System, is even more challenging: hundreds of researchers around the world are currently working on these issues.

The search for exoplanets has triggered, and still causes, great advances in astronomical instrumentation. The largest telescopes on the ground and in space are used to observe them, and many programs are entirely dedicated to their research and study. It is now possible to have an idea of the temperature and chemical composition of the atmosphere of the giant exoplanets, especially the "hot Jupiters" which gravitate close to their stars. However, we are not yet to this point for the less massive planets, which are the ones that could shelter life.

Indeed, the most exciting prospect opened by the discovery of extrasolar planets is the search for planets located neither too close nor too far from their star, where life could develop. Our anthropocentrism suggests that they must have a mass close to that of the Earth, but in reality nothing prevents more massive planets – super-Earths – from carrying life. While planets similar to the Earth are still difficult to discover, we already know a lot of super-Earths, some of which appear to be in favorable conditions. To characterize them and to look for clues about some form of life are some of the main challenges of the coming years. For clear answers, it will probably

Introduction 3

be necessary to wait for the successor of the Hubble Space Telescope and the giant telescopes currently under construction. In parallel, we continue to look for the presence (or absence) of a primitive life present or past on Mars and on certain satellites of the Solar System. Finally, perhaps one day we will communicate with advanced civilizations on distant planets.

Taking stock of such a moving subject is a challenge that we have tried to address. Indeed, it seemed useful to give the bases for understanding the current works on exoplanets, which are so numerous and so diverse that it is easy to get lost. We are aware, however, that if we had waited a few more years to write our book, it would probably have been quite different as surprising discoveries occur at a frantic pace. But is not this also true in many fields of science?

We thank Joanna Kubar, Michel Le Bellac and Michèle Leduc for their careful reading of the manuscript of this book.

Chapter 1
The prehistory of exoplanets

First ideas and speculations

The idea that there might be planets around other stars than the Sun is far from obvious and took a long time to come along. For this, it was necessary to realize that the stars are other suns, and to imagine the presence of planets around these stars. Then, there was only one step, easy to cross, to think that these planets could carry inhabitants. We shall not be surprised to find that the first one who had such ideas, or at least who left written traces of them, was Epicurus (ca. 342 BC–270 BC), a Greek philosopher whose independence of mind is well known. He wrote in his letter to a certain Herodotus:

> "The Universe is infinite ... There is an infinity of worlds, some like this one, others different."

In his *De Rerum Natura*, Lucretius (ca. 94 BC–ca. 54 BC) elaborated the thought of Epicurus. He asserted that the atoms that formed the Earth and the living species could not fail to generate similar systems elsewhere. So,

> "The sky, the earth, the sun, the moon, the sea, all the bodies, in short, are not unique, but rather infinite in number."

It was not possible for Christians to adopt such unorthodox ideas, and one had to await the Renaissance for some bold minds to return to them. Curiously enough, one of the first to be traced is a very influential cardinal, Nicolas de Cuze (or Cues, 1401–1464). In his *De docte ignorantia* of 1440, in which he expressed his skepticism about doctrinal knowledge, he wrote that the Universe was an infinite sphere whose center was everywhere and the circumference nowhere: the Earth had no reason for being its center and, like all celestial objects, could not be motionless, an idea that influenced Copernicus. Giordano Bruno (1548–1600) went much further in writing his *La Cena de le Ceneri* (The ashes banquet), published in 1584. He was a convinced Copernican, thinking that the planets revolve around the Sun, and that others must surround the stars, which he considered like other suns; they could even have inhabitants.

Despite the sad fate of Giordano Bruno, condemned to the stake in 1600 for his heretical ideas, many authors speculated after him. Johannes Kepler (1571–1630) wrote between 1620 and 1630 his *Somnium*, which was published in 1634, after his death, under the title of *Somnium, seu opus posthumum de astronomia lunari* (The Dream, posthumous opus on Lunar astronomy). It is a book of science fiction before the letter, where he considered life on the Moon, which was to be very different from the terrestrial life because the conditions were not the same. In England, the founder of the Royal Society, John Wilkins (1614–1672), published in 1638 *The Discovery of a World in the Moone*. The same year was published the posthumous work of Bishop Francis Goldwin (1562–1633), *The Man in the Moone*. In France, one finds among other writings the *Histoire comique des Estats et Empires de la Lune* (Comic History of the Estates and Empires of the Moon), and the *Histoire comique des Estats et Empires du Soleil* (Comic History of the Estates and Empires of the Sun) by Savinian Cyrano de Bergerac (1619–1655), published posthumously in 1657 and 1662 respectively. In 1752 Voltaire (1694–1778) published his *Micromegas*. All these works are of pure imagination or are philosophical tales.

These books are limited to the Solar System. But in a wider context, René Descartes (1596–1650) constructed at the time a new system of the World, based on vortices of subtle matter. He might have been inspired by Epicurus, who wrote:

> *"The worlds [...] are formed from the infinity by separating themselves by particular vortices, some larger, others smaller. They destroy themselves, some early, some later, some by a cause, others by another."*

For Descartes, one of these vortices is centered on the Sun and drives the planets, others surround the stars and move their own planetary systems (Figure 1.1). Secondary vortices push the satellites of Jupiter and Saturn. Descartes' theories were to have a great influence in the Netherlands for some decades, and even longer in France, until they were dethroned by the Universal gravitation of Newton and its consequences on the motion of the planets.

Bernard le Bovier de Fontenelle (1657–1757), a distinguished scientist whose longevity was exceptional, and who was for a long time the secretary of the Academy of Sciences, was a convinced Cartesian and stayed it until his death, since he still published in 1752 a *Théorie des tourbillons cartésiens avec des réflexions sur l'attraction* (Theory of Cartesian vortices with reflections on attraction). Fontenelle certainly knew the book of Wilkins, which was translated into French in 1656. In 1686, he published his *Entretiens sur la pluralité des mondes* (Conversations on the Plurality of Worlds), which was often reprinted. He is was excellent popularizer. To draw the attention of the reader, he wrote in the introduction:

> *"It seems that nothing should interest us more than to know how the world we inhabit is made, if there are other worlds that are similar to it, and which ones are inhabited as well as ours."*

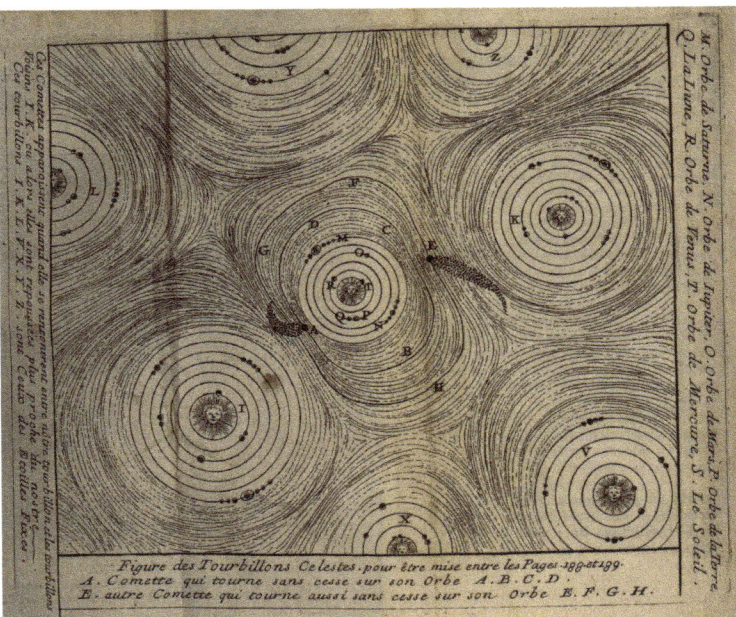

FIGURE 1.1 – For Descartes, planetary systems are born in vortices. Descartes described only the formation of the Solar System, but it is clear in reading his work that he thought that planetary systems have been formed around other stars. His successors, like Nicolas Bion (1652–1733) to whom we borrow this engraving, represented them explicitly. Library of the Paris Observatory.

Fontenelle imagined what the inhabitants of the Moon, Mercury, and Venus might be. So as not to offend the theologians, he supposed that they were different from humans. Oddly enough, he did not care about Mars, which did not seem habitable to him. On the other hand, the satellites of Jupiter and Saturn seemed worthy of being inhabited, and could be "colonies" of their respective planets. Finally, like Giordano Bruno and Descartes, he proposed:

> *"The fixed stars are so many suns, our Sun is the center of a whirlwind that revolves around it, so why each fixed Star could not also*

be the center of a whirlwind that would turn around it? Our Sun has planets that it illuminates, so why each fixed star would not have planets that it would illuminate?"

Following Bruno and Fontenelle, Christiaan Huygens (1629–1695) was ready to admit the presence of living beings on other planets than the Earth. In 1690, he began an important work, the *Cosmotheoros*, which was not published until 1698, shortly after his death. Written with a real concern for popularization, this book is nevertheless in Latin, and therefore destined to a cultured public. One of its translations into French, published in 1724, specifies that it was a work *"in the style of that of M. de Fontenelle on the same subject, but where, for philosophical reasons, conjectures that he proposed only as a simple game of mind are demonstrated to be quite probable."* It is clear that the problem of the plurality of worlds was now considered in a most serious way. But it was for the moment only the outcome of an overflowing imagination, which supplemented the lack of knowledge about the possibilities of life in the Universe. And this is what really interests the public!

After the fantastic projections of Kepler, Fontenelle and Huygens, the subject of the plurality of inhabited worlds hardly progressed, and for good reason: no new observation came to nourish the subject. William Herschel (1738–1822) speculated in 1795 on the habitability of the Sun, but he remained isolated.

Moreover, the time was now for rationalism and the problem of the existence of other planets and their inhabitants was dormant. As it is obviously difficult to demonstrate this existence, the attempts to detect it were somewhat despised by professional astronomers. Camille Flammarion's (1842–1925) essays to revive the subject, the first of which is *La pluralité des mondes habités* (The plurality of inhabited worlds) of 1862, if they excited the public, were badly considered by astronomers. This did not prevent speculation to continue, culminating with the famous formula proposed by Frank D. Drake in 1960 to express the probability of existence of life on other planets (we will talk about it in chapter 12), then with the American *Star Trek* series of 1966, the first to involve exoplanets.

The evolution of concepts on the formation of the Solar System

If Descartes was one of the first to attempt to construct a coherent theory of the formation of the Sun and its cortege of planets, his cosmogony did not rest on any reasonable physical basis, something that Newton did not fail to notice. Yet it had the merit of conceiving that the Universe may be in evolution, a remarkably modern conception although Epicurus was a precursor for it; it foreshadows the future evolutionist cosmogonic models. We can oppose to it the catastrophic cosmogonies like that of Buffon (1707–1788).

For the latter, the Sun and the comets preexisted to the planets, which were born together of a catastrophe: the encounter of a comet with the Sun. Then everything stabilized. The evolutionist and the catastrophic models, which existed before Buffon, clashed for a long time on the basis of arguments generally more ideological than scientific.

The philosopher Emmanuel Kant (1724–1804) adopted, for his part, an attitude that recalls that of Descartes, but he benefitted from the progress of physics. In his *Allgemeine Naturgeschichte und Theorie der Himmels* (Universal history of Nature and Theory of the Heavens, 1755), he proposed that all the matter which formed the Solar System was originally in the form of a great nebular mass. It would have collapsed into a flat rotating disk under the effect of its own gravity, and then was distributed between the Sun and the planets, which were born of inhomogeneities in the disk. The idea of the rotation of the disk comes, of course, from the fact that all the planets rotate in the same direction in orbits located in approximately the same plane, and it is necessary for this that the primordial cloud was itself in rotation.

Pierre Simon Laplace (1749–1813) had ideas similar to those of Kant (did he know the latter?). He exposed and perfected them in the successive editions of his *Exposition du système du monde* (Exposition of the System of the World). In his description of the Solar System, he included the "small planets", now called asteroids, the first of which, Ceres, was discovered on 1 January 1801 by Giuseppe Piazzi (1746–1826). They filled a void in the distribution of the distances of the planets to the Sun, which appeared to Laplace to be a confirmation of the idea that dominated his work: the present disposition of the Solar System resulted from its origin, and owed nothing to chance. Unlike Kant, Laplace could know the observations of William Herschel (1738–1822), who observed and drew many nebulosities, more or less elliptical, each surrounding a star, which he called *planetary nebulae* (Figure 1.2). Herschel and Laplace thought that they were remnants of the primitive nebula which had just formed the central star, but not yet the planets. Both were mistaken, for we know today that the great majority of these objects are the product of a star at the end of its life, which has ejected a part of its matter in the form of a gas, which it illuminates. Nevertheless, the term "planetary nebulae" has been preserved for them.

The "nebular hypothesis" of Kant and Laplace was to dominate the whole of the 19th century, and is still valid in a somewhat modified form. It presents, however, an important difficulty, which seems to have been pointed for the first time in 1861 by Jacques Babinet (1794–1872): how is it that the Sun turns so slowly on itself, in a little more than twenty-five days, while the planets have a much greater kinetic momentum of rotation in their revolution around the Sun? It is necessary that the rotation of the proto-Sun has been slowed in some way. It was not until the 1980s that the problem was solved, in an unexpected way: see Chapter 6.

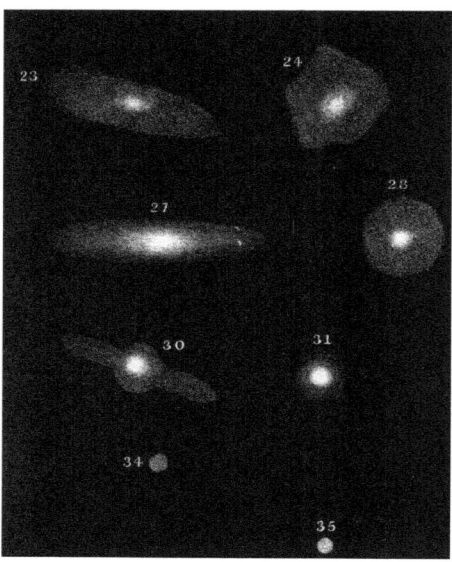

FIGURE 1.2 – Some planetary nebulae drawn by Herschel, with their central star. Library of the Paris Observatory.

Meanwhile, this problem led to a temporary revival of the catastrophic theories: the astronomer Forest Ray Moulton (1872–1952) and the geologist Thomas Chamberlain (1843–1928), estimated in 1906 that the revolution of the planets around the Sun implied that the planets were made of solar matter torn away by the attraction of another star passing nearby, which would have given to this matter the observed velocity. This idea was to be very popular with famous scientists. However, if it were true, planetary systems like ours should be extremely rare, since the passage of a star very close to another can only be exceptional, in view of the immense distances that separate the stars. Now we know that planetary systems are very common. Even before this observation, the new catastrophic theory collapsed when subjected to detailed analyzes, notably that of Lyman Spitzer (1914–1997) in 1939: we therefore returned to Laplace and even temporarily, in a modified form, to the whirlwinds of Descartes.

These vortices were indeed reintroduced by several renowned astronomers, particularly in 1944 by the German scientist Carl Friedrich von Weiszäcker (1912–2007). The latter imagined that in the protoplanetary disk there were vortices, the system of which recalled concentric ball bearings. The planets would not have formed within the vortices as in the model of Descartes, but in the spaces that separated them. This theory did not resist criticism for long: in 1948, the author himself acknowledged that it had to be abandoned. So we came back once again to Kant and Laplace.

Two models now clashed in the 1960s. In the first one, mainly due to the Canadian Alastair G.W. Cameron (1925–2005), the protoplanetary disk surrounding the nascent Sun was supposed to have a mass comparable to that of the Sun. The planets would have formed directly in this disk by gravitational collapse. A large fraction of what would remain from the disk would have been swept by the intense wind of ionized particles from the forming Sun, while the rest would have been captured by the Sun.

In the other model, due to the Russian Viktor Safronov (1917–1999), the mass of the disk would have been only one hundredth of that of the proto-Sun. The disk would have cooled, the dust it contained would have accumulated in the median plane, then agglomerated by the play of molecular forces in small bodies of kilometric size, the planetesimals. Thanks to the mutual gravitational attractions, the planetesimals would have merged into more massive objects, the planetoids. All these objects rotating together in the disc, their relative velocities were small enough to allow their adhesion during their encounters. Later, the most massive fragments could grow by gravity grabbing the surrounding matter. At a great distance from the Sun, well beyond the orbit of the future Earth, the temperature was low enough that most chemical elements apart from hydrogen and helium (oxygen, carbon, nitrogen...) were in the form of ice covering the grains of dust, and were then included in the planetesimals and planetoids. Large nuclei of rocks and ice were able to form, capable of capturing by gravity the surrounding gas essentially composed of hydrogen and helium: the giant planets were born. Closer to the Sun, on the other hand, the only solids were the refractory ones (silicates, metals...), which were not very abundant. The available material has allowed only the formation of objects of relatively low mass: the terrestrial planets (Mercury, Venus, the Earth and Mars), which are rocky. A diagnosis in support of this scenario is provided by measurements of the chemical elements in the giant planets: these are enriched with respect to hydrogen relative to the values measured in the interstellar medium. This enrichment increases from Jupiter to Neptune, in agreement with the increase in the ratio between the mass of the initial core of ice and heavy elements to that of the envelope dominated by hydrogen. This favors the Safronov model, which was adopted as it was until the early 2000s; but we shall see later that things are much more complex, because the planets are not today at the distance of the Sun where they were formed.

The discovery of protoplanetary disks

In our Solar System, there is not much left of the protoplanetary disk, a large part of which has been used to form the planets. Another part was captured by the proto-Sun or ejected, as proposed by Cameron. Some primitive bodies subsist in the form of asteroids, as well as icy objects like Pluto and cometary nuclei, which lie beyond the orbit of Neptune. Other cometary nuclei were expelled into a much more remote reservoir, the Oort's cloud. But if the Solar System is not an isolated case, we should be able to observe protoplanetary disks around other stars that are forming or have recently formed.

In fact, the first detections of such disks were fortuitous, and were due to the InfraRed Astronomical Satellite (IRAS), a joint project of the United States, the Netherlands and the United Kingdom. IRAS aimed to detect and map the emission of interstellar dust particles. Because of their low temperature, a few tens of degrees above absolute zero, these grains radiate essentially in the infrared. IRAS indeed detected for the first time, in 1983–1984, the thermal radiation of interstellar and interplanetary dust. But it also discovered that several stars that were much older than the stage of planet formation radiated in the infrared: for example, the star Vega (α Lyrae) was found much brighter in the infrared than expected, as were 11 other nearby stars. Although the angular resolution of the telescope on the satellite was insufficient to measure the dimensions of the emitting zone, it appeared that the dust emission was localized to a few tens of astronomical units of the star. The disk could not be the protoplanetary disk itself, since the star was already well formed, but a disk of debris that remained after the formation of planets. A year later, another disk was discovered, this time from ground observations, around a younger star, β Pictoris. It was also a disk of debris (we shall see later that one of the planets that gravitate around this star has been seen directly). In 1986–1987, a disk was discovered around the very young star HL Tauri by ground observations with radio telescopes in the millimeter wave domain; this disk contained gas since an emission line of carbon monoxide CO has been observed, and it is rotating around the star: this time, this was a real protoplanetary disc.

The discoveries of disks around young stars then multiplied, in particular thanks to millimeter observations from the ground, measurements with the Hubble Space Telescope in visible light, and infrared observations by the astronomical satellites ISO of the European Space Agency and Spitzer of NASA. It appears today that at least half of the young stars show an excess of infrared emission betraying the presence of a disk. Some of these disks are presently observed with great detail, as we shall see in Chapter 6.

The first attempts to detect exoplanets

The exoplanets themselves are much harder to discover than the disks. It is not surprising, then, that the nineteenth century and the first half of the twentieth century are dotted with announcements of discoveries of exoplanets that have all proved to be false, which had not encouraged most astronomers to look after the question.

The prehistory of exoplanets

FIGURE 1.3 – Fraunhofer's telescope at the Dorpat Observatory, today Tartu. This instrument, recently restored, is the prototype of all the astronomical telescopes of the nineteenth century. Observatory of Tartu.

The first of these false alarms concerned a planet that would revolve around one of the components of the double star 70 Ophiuchi. William Herschel had announced in 1803 the existence of double stars orbiting around one another,[1] and this subject quickly aroused an important activity on the part of the astronomers. They now had at their disposal excellent refracting telescopes derived from that which was built in 1824 by Joseph von Fraunhofer (1787–1830) for the Russian Imperial Observatory at Dorpat (now Tartu, Estonia, Figure 1.3). It was then possible to systematically

---

[1] More precisely orbiting around their common center of gravity.

observe the revolution of double stars sufficiently separated, hoping to deduce their masses if their distance could be determined. In 1855, the two components of 70 Ophiuchi had almost accomplished a complete revolution around each other since their discovery by Herschel in 1779 (the period is 93 years), but it seemed difficult to represent the orbit by an ellipse. Would there be a third body? This is what an English astronomer who headed the Madras Observatory in India, Captain William S. Jacob (1813–1862), claimed. He thought that the agreement between a theoretical elliptical orbit and the observed one could be improved by adding a third body orbiting in 26 years around the less massive of the two stars. Although his deduction proved to be false, Jacob had the merit of being the first astronomer to have attempted to discover an exoplanet. The history of the 70 Ophiuchi's planet then underwent many twists and turns. Forty years later, a Chicago astronomer, Thomas J.J. See (1866–1962), made new measurements and confirmed the existence of a satellite of the weakest star, but announced a period of 36 years and an orbit differing from that of Jacob. However, another American astronomer, Forest E. Moulton, pointed out that See's system would be gravitationally unstable, and indeed a good elliptic orbit was obtained without the need for an additional object. See got into the huff and sent a furious letter to the *Astronomical Journal* where he had published his so-called discovery in 1895. The result was that he was now forbidden to publish in this journal. The rest of his career was permanently affected.

The story is not over! In 1943, two other astronomers, the Dutch Dirk Reuyl (1906–1972) and the Swedish Erik Holmberg (1908–2000), published a new study of 70 Ophiuchi in the American *Astrophysical Journal*, based on numerous photographic plates, which were supposed to give more precise results than the visual observations of their predecessors. They concluded on the probable existence of an object of about 0.01 solar mass (i.e. 10 times the mass of Jupiter), whose period of revolution would be 17 years, still different from the periods previously announced. This result prompted research on other stars: the same year and the following year, the Danish astronomer Kaj Aa. Strand (1907–2000), using the 61-cm diameter refracting telescope of the Sproul Observatory in Pennsylvania, announced the discovery of four new objects gravitating around the stars ζ Aquarii, μ Draconis, ξ Bootis and 61 Cygni. He confirmed in 1957 his result on this last star. A Dutch astronomer, Peter van de Kamp (1901–1995), who worked at the same place, announced in 1944 the discovery of two similar cases, one of which had already been suspected in 1936 by his cousin Dirk Reuyl as early as 1936. Finally, van de Kamp went after Barnard's star, which was discovered in 1916 by the American astronomer Edward E. Barnard (1857–1923): it is the nearest star to us after α and Proxima Centauri. From 1938 to 1975, no fewer than 4079 photographs of the field of this star were obtained. Van de Kamp deduced the presence of two planets gravitating around the star, somewhat smaller in mass than Jupiter. This would have been of the greatest interest if the subsequent observations had not invalidated this result.

Indeed, none of the previous results has been confirmed, and no planet has yet been discovered by astrometry, except perhaps one (HD 176051 b). The examination of the original publications can only cast doubt on the assertions of their authors: the detections announced are in all cases unclear and to the limits of the possibilities of the instrument. But it is understandable that after the relentlessness of Sproul's astronomers, who had family or friendly ties between them, they have not been able to give up announcing their supposed discoveries which had taken so long, even if they aroused bitter disappointments and sometimes the sarcasm of their colleagues.

In fact, it was practically impossible at the time to detect exoplanets by classical astrometry: their effects are too weak. One had to try other methods. The first to have really understood it is an American astronomer of Russian origin, Otto Struve (1897–1963). Struve was the fourth of a dynasty of illustrious astronomers, who in particular founded and directed the Pulkovo Observatory near St. Petersburg. He emigrated in 1920 to the United States and never returned to Russia. His career was brilliant, but above all he had a great intuition as evidenced by the following extracts from a 1952 article from him:

> "One of the burning questions of astronomy deals with the frequency of planet-like bodies in the galaxy which belongs to stars other than the Sun. [...] I have suggested elsewhere that the absence of rapid axial rotation in all normal solar-type stars [...] suggests that these stars have somehow converted their angular momentum of axial rotation into angular momentum of orbital motion of planets. Hence, there may be many objects of planet-like character in the galaxy.
>
> But how should we proceed to detect them? The method of direct photography used by Strand is, of course, excellent for nearby binary systems, but is quite limited in scope.[2] There seems to be at present no way to discover objects of the mass and size of Jupiter: nor is there much hope that we could discover objects ten times as large as Jupiter, if they are at distances of one or more astronomical units from their parent stars.
>
> But there seems to be no compelling reason why the hypothetical stellar planets should not, in some instances, be much closer to their parent stars than it is the case in the Solar System. It would be of interest to test whether there are any such objects.
>
> We know that stellar companions can exist at very small distances. It is not unreasonable that a planet might exist at a distance of 1/50

---

[2] The fact that all detections using astrometry, such as those of Strand, were invalidated was not yet known at the time.

> astronomical unit, or about 3,000,000 km. Its period around a star of solar mass would then be about 1 day.
>
> [...] Our hypothetical planet would have a velocity of roughly 200 km/sec. If the mass of this planet were equal to that of Jupiter, it would cause the observed radial velocity of the parent star to oscillate within a range of ±0,2 km/sec — a quantity that might be just detectable with the most powerful Coudé spectrographs in existence. A planet ten times the mass of Jupiter would be very easy to detect, since it would cause the observed radial velocity of the star to oscillate with ±2 km/sec.[...]
>
> There would, of course, also be eclipses [by the planet passing in front of the star]. Assuming that the mean density of the planet is five times that of the star (which may be optimistic for such a large planet) the projected eclipsed area is about 1/50th of that of the star, and the loss of light in stellar magnitudes is about 0.02. This, too, should be ascertainable by modern photoelectric methods, though the spectrographic test would probably be more accurate. The advantage of the photometric procedure would be its fainter limiting magnitude compared to that of the high-dispersion spectrographic technique."

Thus, Struve has perfectly established the main techniques which are used today to detect the extrasolar planets, techniques that we will describe in detail later. But it does not seem that his visionary proposals have had much echo in his day. It was generally thought that the possible planetary systems must necessarily resemble the Solar System, in which case a detection would be impossible. Moreover, it seemed that the astrometric method had not said its last word. However, the techniques that should allow the detection were gradually taking place, although they were not intended for exoplanets: in the late 1980s, astronomers were very interested in brown dwarfs, these aborted stars with a mass of between 0.01 and 0.08 solar masses, where the nuclear reactions could not be initiated and which are therefore cooling very slowly. In fact, some brown dwarfs that gravitate around normal stars were found by the radial velocity method. One of them, discovered in 1989 around the solar-type star HD 114762, has a mass equal to or greater than 11 times that of Jupiter, which is at the limit between the brown dwarfs and the planets (we shall see farther than this limit is rather vague). The method has been perfected not only by the use of fixed conventional spectrographs fed by an optical fiber, which improves their stability, and also in two ways that were not foreseen by Struve: by interposing in a high-resolution spectrograph a tank containing gaseous hydrogen fluoride, which gives lines at many wavelengths superimposed on the spectrum of the star, and by the simultaneous use of the very numerous lines of the central star (which requires it to be sufficiently cold), which are correlated with those of a standard spectrum. We will return to this in the next chapter.

At the same time, in 1985, some efforts were made by NASA to detect exoplanets (this word was not yet used). In 1988, NASA convened a group of experts to formulate strategies for the detection and study of other planetary systems. Their report named TOPS (for *Towards Other Planetary Systems*) was published in 1992. However, things unfolded unpredictably, and the recommendations in this report had little effect.

Bibliography

Struve, O. (1952) Proposal for a project of high-precision stellar radial velocity work, *The Observatory* 72, 199, accessible via http://cdsads.u-strasbg.fr/abs/1952Obs....72..199S

Burke, B.F., ed. (1992) *TOPS, Toward Other Planetary Systems: A Report*, accessible via internet

Connes, P. (2020) History of the Plurality of Worlds, edited by J. Lequeux, Springer, New York

Chapter 2
The first detections

Let us place ourselves in the early 1990s. No exoplanet in sight, but in recent years, observations with the infrared satellite IRAS had revealed the existence of protoplanetary disks around young stars, reviving the debate around the formation of stars and their possible planetary systems. The search for hypothetical planets around other stars had lost none of its interest, quite the contrary. However, after the failures of the preceding years, the astronomers realized that the existing astrometric methods did not allow their detection. Then a big change happened, with two discoveries that showed not only that the Solar System was not unique, but also that the variety of exoplanets was much greater than the imagination of astronomers. In 1992, two exoplanets were detected around a star at the end of its life, a pulsar, that is what remains of a star after it has exploded into a supernova. Then in 1995, finally, an exoplanet was discovered in orbit around a star similar to the Sun. It was a giant planet, with a mass at least equal to half that of Jupiter, but incredibly close to its star, at 0.05 astronomical unit, 100 times closer to the star than Jupiter is from the Sun.

An unexpected discovery: planets around a pulsar!

Since the 1970s, radio astronomers have been studying the radio signal of pulsars. These "pulsating stars" at the end of their life are in very fast rotation and emit, as a rotating lighthouse, a radio signal whose period is extremely stable. Astronomers have several reasons for "timing" the radio signal of the pulsars, ranging from the detection of possible gravitational waves to the study of the nature of these neutron stars. If the pulsars are accompanied by planets, their presence should result in disturbances of this signal. After several false alarms, in 1992, the Polish astronomer Alexander Wolszczan, using the Arecibo radio telescope in Puerto Rico, announced the discovery of two exoplanets around the pulsar PSR B1257+12. This is a "millisecond pulsar", so named because of its extremely short rotation period: 0.0062 seconds. The surrounding planets, about four times the mass of the

Earth, orbit at less than one astronomical unit. A third, smaller planet, similar in mass to the Moon and closer to the star, was discovered later.

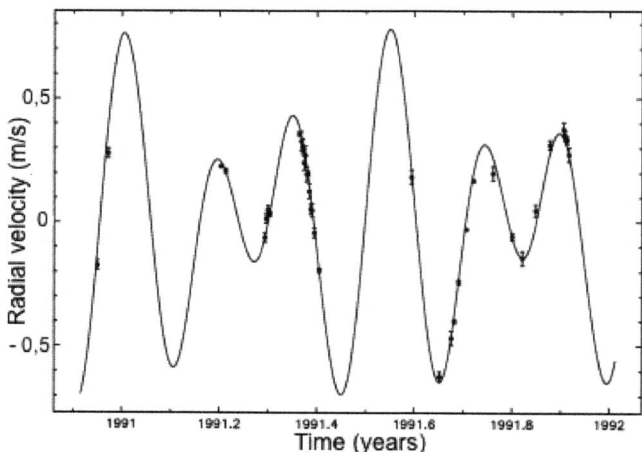

FIGURE 2.1 – The periodic variations of the radial velocity of the pulsar PSR B1257+12 during one year, deduced by the Doppler-Fizeau effect on the the pulsar period, which varies periodically due to the gravitational effect of the planets that surround it. The curve corresponds to the disturbances produced by the presence of two planets, the points of measurement are indicated with error bars. The refinement of the measurement of the period variations of the pulsar made it possible to detect not only the third planet but also the very small effect on this period of the mutual perturbations of the two first ones which slightly modify their movement. Their masses, 4.3 and 3.9 Earth masses, have been determined in this way (they are thus known as "super-Earths") and the inclination i of their orbits in the plane of the sky is about 50° (see Figure 2.2). According to Wolszczan, A. (2012) *New Astronomy Reviews* 56, 2.

The discovery of these first exoplanets made a great noise! It was also a great surprise, as millisecond pulsars are thought to result from the evolution of a tight pair of a normal star and a neutron star, the latter resulting from the explosion of supernova. If there were planets around this star, it is unlikely that they resisted the explosion. We can imagine, therefore, that the normal star, as it aged, saw its outer envelope gradually falling by spiral on the neutron star, whose rotation accelerated and which became a millisecond pulsar. The existence of planets around the millisecond pulsar probably involves the preliminary formation of a protoplanetary disk, which could be an accretion disk made of material from the other star, but this evolution remains poorly understood. The exoplanets around pulsars seem therefore uncommon, although about twenty have been discovered at the time of wrinting. Whatever their formation scenario, they are unlikely to resemble the planets of the Solar System. Nevertheless, they do exist and their discovery was a major step in the search for extrasolar planets.

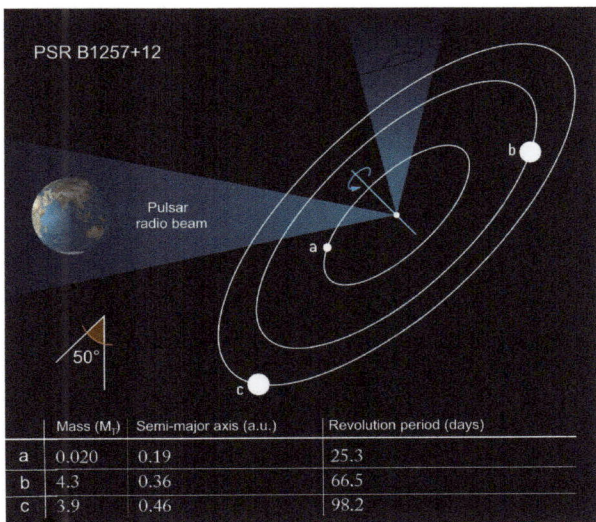

FIGURE 2.2 – The planetary system of the pulsar PSR B1257+12. The characteristics of planets a, b and c are indicated: mass (in Earth masses), semi-major axis and period of revolution. Their orbits are approximately in the same plane, inclined by about 50° with respect to the direction of the pulsar as seen from the observer. The Earth is swept by the radio pencil beam emitted from the pulsar, whose axis of rotation is probably perpendicular to the plane of the planet's orbits, as shown in the figure. If this is correct, the radio beam is emitted at 50° from the axis of rotation of the pulsar. Adapted from Casoli, F. & Encrenaz, T. (2005), Planètes extrasolaires, Paris, Belin.

1995: The first planet around a star like ours!

To finally detect planets around stars like the Sun, in the middle of their lives, astronomers have turned to velocimetry, a method that has proved its value in the study of double stars. What is it about? As in the case of astrometry, the principle is to detect the effect of the presence of a planet on the motion of the star: a star accompanied by one or several planets has its motion slightly perturbed in a periodic way. Astrometry tries to measure this movement on the celestial sphere, with respect to the neighboring stars. Velocimetry, on the other hand, measures the variations of the radial velocity of the star with respect to the terrestrial observer using the Doppler-Fizeau effect (Box 1).

Box 1. Detection of a planet by the Doppler effect (or more correctly the Doppler-Fizeau effect)

Consider a star accompanied by a planet. The orbit of this planet is an ellipse; but the star itself describes a small ellipse with respect to the mass center of the system, which is generally close to the center of the star (Figure 2.3).

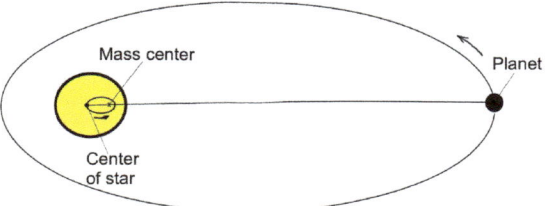

FIGURE 2.3 – Diagram showing the motion of a star and its planet around the center of gravity of the system. The center of the star describes an ellipse which is the mirror image of the planetary orbit, reduced by a factor equal to the ratio of the masses. The center of gravity, fixed, is the common focus of these ellipses. Diagram by the authors.

The astrometric method for detecting the planets tries to observe the projection of this second ellipse in the plane of the sky, which is the plane perpendicular to the star-observer direction.

The velocimetric method measures the variation of the star velocity along the star-observer line of sight, using the Doppler-Fizeau effect. In the same way as the sound of a car approaching us seems to be higher, while the noise of a car moving farther away seems to be lower (as is also the case with the radio waves of the radars controlling the speed of cars), the emission lines present in the spectrum of the star are shifted towards the red when the star moves away from us, and towards the blue when it approaches us. This variation in wavelength is equal to the variation in the radial velocity of the star divided by the velocity of light.

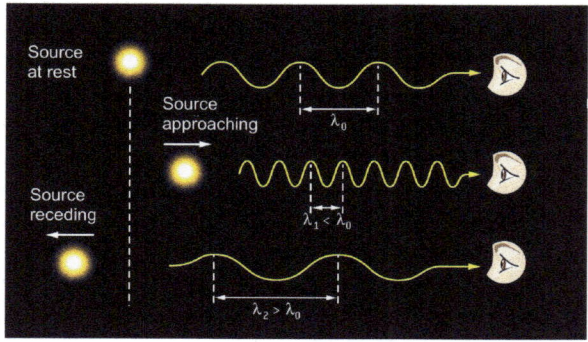

FIGURE 2.4 – The Doppler-Fizeau effect. Adapted from Casoli, F. & Encrenaz, T. (2005) *Extrasolar planets*, Paris, Belin.

> In general, the orbit of the planet, and that of the star, is inclined to the plane of the sky, and the astrometric and velocimetric methods are complementary if both can be used. There are, however, exceptions: if the star-planet system rotates in the plane of the sky, its axis of rotation being aligned with the line of sight, astrometry measures a circle or an ellipse on the sky, but the velocimetry method does not measure any variation of the radial component of the velocity. The opposite case is that where the Earth is in the plane of the planetary orbit; planetary transits are then observed, the planet passing alternately in front of and behind its star.

The great advantage of velocimetry is that it allows, for the time being, a much better accuracy than astrometry. In the 1990s, it was already possible to measure the radial velocity of stars with an accuracy of about ten meters per second. In the case of the Solar System, the motion of Jupiter induces a variation of 12.5 m/s in the velocity of the Sun with its period of revolution of 11.86 years, so that it should have been be possible, if one had much patience, to find a Jupiter around a nearby star. On the other hand, detecting an Earth at 1 a.u. of its star, which would give only a variation of velocity of 9 cm/s, was and remains impossible. To make the method successful appeared to require long-term programs, a lot of observation time and a spectrometer that is perfectly stable in the long term: for detecting a Jupiter, we should observe for at least a dozen years!

These are long-term programs in which several teams engaged in the early 1990s, searching systematically for both companion stars and massive exoplanets. The researchers who carried out these projects were armed with patience because, in the absence of other information, they expected that these giant planets, if they existed, would have, like ours, periods of revolution of about ten years, or even more! Since the programs were to extend for several years and therefore required long hours of observation, they used middle-class telescopes dedicated to this work, equipped with high-performance spectrographs operating in a wide range of visible light spectrum, capable of simultaneously recording several hundred spectral lines. These spectrographs correlate the observed spectrum with a standard spectrum to measure the spectral shift by taking advantage of the gain of precision allowed by the simultaneous observation of many lines. This works better if the star is colder, because its spectrum then comprises a greater number of lines. The wavelength calibration is performed in Europe by the simultaneous measurement of the spectrum of thorium. In the United States, it is preferred to superimpose on the observed spectrum the spectrum of iodine or hydrogen fluoride, gases which produces numerous lines, in order to have a wavelength calibration in the spectrum itself. The spectrographs, which must have a great stability for many years, are in a fixed position, placed on the ground in thermostatically controlled enclosures, and the light arrives from the telescope by an optical fiber.

What information can astronomers deduce from such an observation? Let us assume, for simplicity, that the planet is in circular orbit around the star. The curve described by the radial velocity of the star with respect to us is a sinusoid whose period is that of the revolution of the planet around the star. To observe a sinusoidal period unambiguously, we must therefore observe at least as long as the period of revolution. We can show that the amplitude of variation of the radial velocity of the star is proportional to the mass of the exoplanet, and inversely proportional to the square root of the distance from the planet to the star (Box 2): the effect is therefore maximum for a massive planet, or in orbit close to its star, or both.

Astronomers who engaged in velocimetry programs in the early 1990s were confronted with the following problem: they could theoretically detect the motions induced by a giant planet comparable to Jupiter, but this detection might take about ten years; the detection of a terrestrial planet would take only a year, but the stellar motion induced by such a planet is too small given the performance of the spectrographs of the time. It is here that chance smiled at them. While they imagined the giant exoplanets far from their stars, as we observe in the Solar System, the planetary systems to be discovered are very different: there are many giant exoplanets located in the immediate vicinity of their stars! It is this totally unforeseen fact (except by Otto Struve, whose prediction was forgotten, see the previous chapter), which allowed, in 1995, the first discovery of a planet around a solar type star.

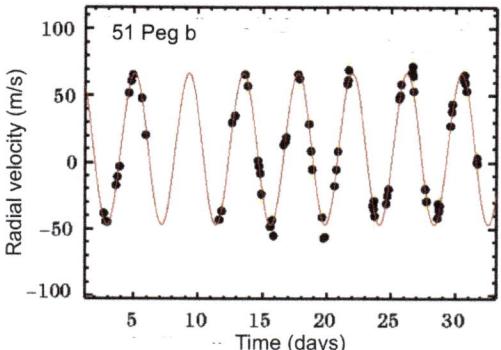

FIGURE 2.5 – The discovery of the exoplanet around the star 51 Peg, from multiple observations indicating a periodic signal. The periodicity of the curve corresponds to the period of revolution of the exoplanet around the star. Note the scale of the radial velocity variation, which is much greater than in the case of the millisecond pulsar (Fig. 2.1). According to Mayor, M. & Queloz, D. (1995).

The first detections 25

The official announcement of this discovery was made in Florence on October 6, 1995, at an international conference on stellar physics. The authors are Michel Mayor and Didier Queloz, from the Geneva Observatory. For a year, they have collected measurements using a high resolution spectrometer, ELODIE, installed on the 1.93 m diameter telescope of the Haute-Provence Observatory (Fig. 2.6). The first exoplanet detected around a solar-type star, 51 Peg b[3], had very surprising characteristics: with a mass of at least half the one of Jupiter, it revolved around its star in a quasi-circular orbit with an incredibly short period of only four days. The average distance to the star was therefore only 0.05 astronomical units! The news had the effect of a bomb. First of all, this discovery finally confirmed what was then only an intuition or a supposition: the Solar System is not unique in the Universe. But, moreover, this exoplanet – a giant exoplanet very close to its star – was very different from the planets of our Solar System! The latter, though not unique, is not a universal model, and it will soon be seen that "hot Jupiters", like 51 Peg b, are relatively frequent.

FIGURE 2.6 – The 1.93 m telescope of the Observatory of Haute-Provence, which allowed the detection of the first exoplanet around a solar type star. © Observatoire de Haute-Provence.

The implications of this discovery are very important. Let us recall that in the formation scenario commonly accepted for the Solar System, giant planets form far from the proto-star because the temperature is low enough to allow the formation of massive nuclei, which then capture by

---

[3] 51 Peg b is the first exoplanet discovered around star number 51 of the Pegasus constellation, the star itself being regarded as 51 Peg a. See Appendix 4 for how to name stars and exoplanets.

gravity the surrounding gas. The formation model of the newly discovered stellar system must therefore differ from the one we know. How can this be explained? As soon as their discovery was announced, Michel Mayor and Didier Queloz proposed a scenario that they developed in their article published on November 23, 1995, *A Jupiter-mass companion to a solar-type star*: the giant planet probably formed far from its star, but was then able to migrate into the system as a result of its interaction with the protoplanetary disk. The basis of this theory, subsequently developed by many authors, is now widely accepted by the scientific community. We will discuss this in chapter 7.

Box 2. The mass and dimensions of the orbit of an exoplanet detected by velocimetry.

In the general case where the orbit of the exoplanet is inclined by the angle i on the plane of the sky, the amplitude Δv of the variation of velocity of the star in the case of a circular orbit is written as follows (M is the mass of the star and m that of the planet, assumed to be very small before that of the star; T is the period of revolution):

$\Delta v = (2\pi G)^{1/3}\ T^{-1/3}\ M^{-2/3}\ m\ \sin i,$

or, in numerical units that are more relevant:

$\Delta v = 28.4\ \text{m/s}\ (T/\text{yr})^{-1/3}\ (M/M_{\text{Sun}})^{-2/3}\ (m\ \sin i)/M_{\text{Jupiter}}$

This means that if we observe a star of the same mass as the Sun, accompanied by a planet with the mass of Jupiter, but orbiting at the same distance as the Earth, we would see a variation in the velocity of this star with a period of one year; this variation could be up to 28.4 m/s if the inclination of the system is favorable, i.e. if it is seen edge-on.

Knowing the period T, we can obtain the quantity $m \sin i$ if we can determine the mass of the star, which is generally possible from the characteristics of its spectrum (its spectral type). As the orbital inclination is not usually known, a lower limit of the mass of the planet is thus obtained. There is, however, one exception: if one observes passages of the planet in front of the star (chapter 3), the observer is practically in the plane of the orbit and i is very close to 90°: the mass of the planet is then known.

There is another important information that can be deduced from the period of revolution of the planet and the mass of the star: the radius a of the orbit is given by the third law of Kepler:

$(a/\text{a.u.})^3 = (M/M_{\text{Sun}})\ (T/\text{yr})^2$

For the more general case of an elliptic orbit with eccentricity e, $m \sin i$ must be replaced by $m \sin i\ (1 - e^2)^{-1/2}$. The eccentricity can be deduced from the shape of the curve which gives the velocity as a function of time (Figure 2.7).

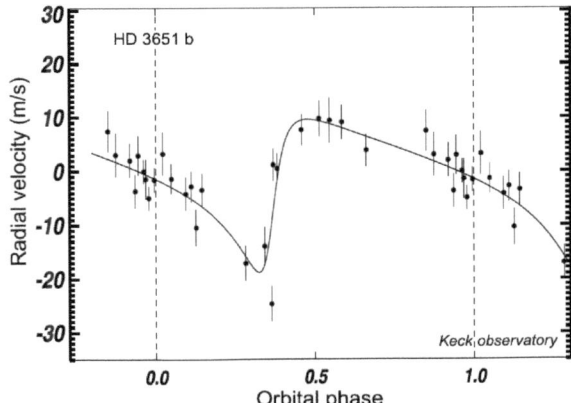

FIGURE 2.7 – An example of a velocity curve for a very eccentric orbit of the planet around its star. The planet HD 3651 b has a period of revolution of 62.23 days and a mass greater than 0.20 $M_{Jupiter}$; its orbit has a half-axis of 0.284 a.u. and an eccentricity of 0.63. For a circular orbit, the curve would be sinusoidal (see an example in Figure 2.5). From Fischer D.A. et al. (2003) Astrophysical Journal 590, 1081.

Very quickly, the discovery was confirmed by independent measurements carried out by other teams. The first was by Geoffrey Marcy of the University of California at Berkeley and Paul Butler of the Carnegie Institution of Washington, who announced their results at the annual conference of the American Astronomical Society in December 1995. In the wake, they announced the possible detection of another companion of 51 Pegasi, more massive and more distant, but they later infirmed this result. Most importantly, they reported in June 1996, the discovery of two other exoplanets. Located around the stars 47 Ursae Majoris and 70 Virginis and having masses at least equal to 2.4 and 6.6 Jupiter masses respectively, these planets are at average distances of 0.43 and 2.1 a.u. from their stars. Here are two new exoplanets that do not meet the formation criteria of the planets of the Solar System. In the following months, discoveries multiplied; in most cases, the exoplanets found were very massive, and very close to their star.

The success of velocimetry

Until the end of the 20th century, velocimetry was the only method allowing the exploration of exoplanets. We have seen that this is an extremely powerful method: from the simple observation of the variation in the velocity of the star, we can deduce not only the presence of one or more planets, but also the mass of this planet (more precisely a lower limit), its period of revolution around the star, the shape of its orbit, circular or elongated. A hundred exoplanets were detected in the five years following the announcement of the first discovery, most of them around solar-type stars. Several systematic researches were carried out and still continue, using dedicated spectrographs that are increasingly performing, notably HIRES, with the Keck telescope of ten meters in diameter in Hawaii, and HARPS, with the ESO 3.60-meter telescope in Chile (Figure 2.8), which achieve measurement accuracies better than 1 m/s.

FIGURE 2.8 – The HARPS spectrograph of the Southern European Observatory, mounted on the 3.60 m diameter telescope of La Silla (Chile). It is enclosed in a vacuum container, here open. One can see the large grating that disperses light, in two parts, whose total dimensions are 200 × 800 mm. © European Southern Observatory.

Beginning in 2000, another technique started to develop: the transit method, which detects an exoplanet that passes in front of its star by the small decrease in its light, which can be measured at each passage (see Chapter 3). The velocimetry then becomes essential to validate the planet candidates discovered by the observations of transits: indeed, it is necessary to ensure that the temporary decrease in the luminosity of the star at the time of transit is not an artifact due to a periodic fluctuation of the star itself. Following the growing success of the transit technique, first from the ground and then in space with the CoRoT and Kepler satellites, validation campaigns by measuring radial velocities were thus set up. Of the approximately 4270 exoplanets discovered at the time of writing, about 885 were detected by velocimetry; the majority of the others were discovered by transit and, whenever possible, confirmed by velocimetry.

The improved performance of instruments (most of them are now spectrometers in vacuum cooled to a very low temperature to achieve maximum stability) now makes it possible to obtain, in radial velocity, limits of a few tens of cm/s. Thus, radial velocity measurements make it possible to reach mass limits that are increasingly close to the Earth's mass (Fig. 2.9). In the future, further progress is expected with the ESPRESSO instrument, derived from HARPS, which is presently being mounted at the ESO's Very Large Telescope and whose accuracy is expected to be ten times better. Even more powerful instruments are being studied to equip future large telescopes, the Extremely Large Telescopes (ELTs): the 39-meter E-ELT in construction by ESO at Cerro Armazones in Chile and the two US projects, the TMT (Thirty Meter Telescope) at Hawaii and the 21-meter Giant Magellan Telescope (GMT) at Las Campanas, Chile.

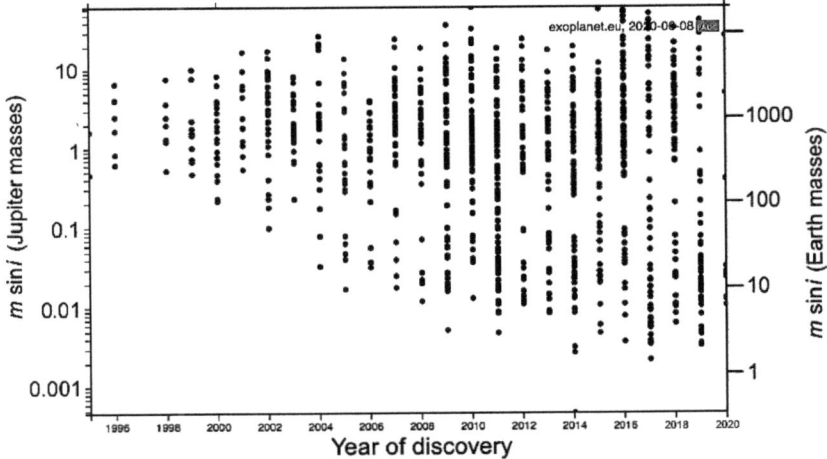

FIGURE 2.9 – Minimum value of the mass of exoplanets discovered by velocimetry, according to their year of discovery. The vertical scale is graduated into masses of Jupiter (left) and terrestrial masses (right). The smallest planets detected by velocimetry have masses close to that of the Earth. From http://exoplanets.eu/.

Velocimetry has broadened its field of research in several directions. First, the targets, which were initially solar-type stars, are today other types of stars, especially the less massive ones (K and M dwarfs), which represent a much larger population than the previous ones. These stars, cooler than the Sun, radiate most of their energy in the near infrared. To study them, it was therefore necessary to design high-resolution spectrographs operating in this wavelength range. Such instruments are already operating in the United States at the Mount Wilson and Mount Palomar observatories; the SPIRou instrument is presently installed at the Canada-France-Hawaii telescope. The interest of studying these stars is that it is easier to discover small planets around them, the radial velocity variation induced by the presence of a planet depending on the ratio of the mass of the planet to that of the star (Box 2): it is therefore easier to detect a terrestrial planet around a small star than around a solar-type star.

Finally, the improvement of the velocimetry technique has allowed the discovery of planetary systems comprising several planets. When a first companion is detected around a star, it is possible to extract from the observed signal the periodic function corresponding to its motion and to detect, from the residual curve, the presence of one or more companions in the system. The first discovery of a multiple planetary system by velocimetry was made by Paul Butler and his colleagues in 1999 around the star υ Andromedae (Figure 2.10). Today, observations show that nearly 25% of all planetary systems detected are multiple systems. It is certain that this number will increase as more and longer campaigns are able to detect more and more distant planets, similar to Saturn or Uranus in our Solar System.

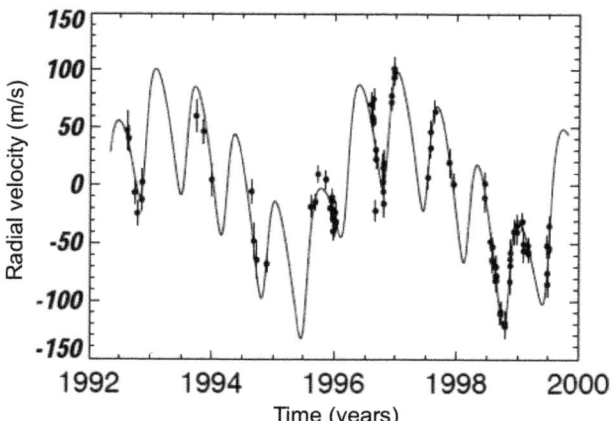

FIGURE 2.10 – The radial velocity curve of the star υ Andromedae, after subtraction of the effect of a planet very close to the star, which produces a sinusoidal signal of period 46,171 days with an amplitude 75 m/s. It shows the existence of two other planets, whose periods of revolution are 241 and 1267 days respectively. From Butler, R.P. et al. (1999).

Bibliography

Mayor, M. & Queloz, D. (1995) A Jupiter-mass companion to a solar-type star, *Nature* 378, 355

Butler, R.P., et al. (1999) Evidence for multiple companions of Andromedae, *Astrophysical Journal* 526, 916, http://cdsads.u-strasbg.fr/abs/1999ApJ..526..916B

Wolszczan, A. (2012) Discovery of pulsar planets, *New Astronomy Reviews* 56, 2

The site http://exoplanet.eu/ contains the list of exoplanets constantly updated, with details on exoplanets and their host stars and extensive bibliographical references.

Chapter 3
The method of transits

What is a planetary transit?

On the eve of the twenty-first century, measurements of radial velocities on nearby stars, mostly solar-type ones, have detected hundreds of exoplanets, most of which have masses comparable to those of Jupiter. However, as we have seen, the velocimetry method provides only a lower limit of the mass of the planet, because the angle under which the system is observed is unknown.

Another method is going to remove this ambiguity: the method of transits. The principle is very simple: if a planet passes in front of its star (in the manner of Venus passing in front of the solar disc in 2004 and 2012), the overall flux of the star is slightly diminished (Figure 3.1). The decrease in flux depends on the relative size of the planetary disk relative to the stellar disk. If an observer external to the Solar System observed a transit of Jupiter in front of the Sun, the diameter of Jupiter being about one tenth of the solar diameter, the solar flux would be reduced by the square of the ratio of the diameters, i.e. about 1%. In the case of the Earth, whose diameter is ten times smaller than that of Jupiter, the decrease would be only 0.01%. The techniques of stellar photometry make it possible to obtain from terrestrial observatories an accuracy of the order of 0.1%, which makes it possible to detect giant exoplanets by transit from the Earth. On the other hand, the detection of exoplanets of the terrestrial type – the "exo-Earths" – cannot be envisaged with terrestrial telescopes, unless the star is much smaller than the Sun: observation from space is required to obtain the stability needed for the required photometric precision. Two space missions, CoRoT and then Kepler, have been implemented with this objective in mind.

To observe a planetary transit, it is therefore necessary to select a star and to measure its flux continuously with the greatest possible precision. From the Earth, the continuity of observations poses a practical problem, linked to the rotation of the Earth on itself. It is therefore advisable to observe during the winter from a high latitude site (Antarctica for example), so as to benefit from several months of uninterrupted observations, or to mobilize a network of telescopes, which also makes it possible to minimize the interruptions of observation in case of bad weather on one of the sites.

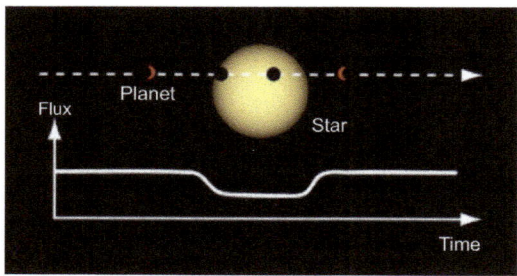

FIGURE 3.1 – Schematic of a planetary transit. Diagram by the authors.

What can we learn from observing a planetary transit? First, the period of revolution of the planet around its star is given by the temporal deviation between two successive decreases of the signal. Then, the relative decrease of the stellar flux corresponds to the occultation presented by the planet on the stellar disk, and is therefore related to the ratio of the radii as $[1-(R_P/R_*)^2]$, R_P being the radius of the planet and R_* that of the star. If one can estimate the radius of the star, which is possible if one knows its distance and its luminosity, as is generally the case, one immediately deduces that of the planet. Moreover, the angle of inclination i of the orbit with respect to the plane of the sky is very close to 90°, since the planet passes in front of its star: hence, by measuring the variation of the radial velocity which gives the product $m \sin i$, we obtain its mass m itself. From the mass and the radius, the density of the planet is known, which gives access to its physical characteristics: gaseous, icy or rocky. However, we will see later that knowing the density is not always sufficient for an unambiguous characterization.

Of course, a condition is necessary for the transit to be observable: the planet must pass in front of the star. Here a very favorable element is to be considered: many giant exoplanets are close to their star, so the probability of transit is much higher than it would be for a system similar to the Solar System. For a Jupiter-like exoplanet located at 0.05 a.u. of a solar-type star, the probability of transit is of the order of 10%. It is therefore not surprising that in the late 1990s astronomers undertook systematic transit research in front of stars that were hoped to host a "hot Jupiter".

In 1999, the first exoplanet transit was announced, that of HD 209458 b. This is a giant exoplanet orbiting close to a very bright nearby solar type star, making it an ideal target for transit studies. The observations were first conducted by Gregory Henry and then independently by David Charbonneau and his team. The transit curve shows a signal decrease of 1.5% during two and a half hours (Figure 3.2). This result confirmed that transits of giant exoplanets are observable from the Earth. Figure 3.3 shows an example of multiple transits observed recently from the European Southern Observatory.

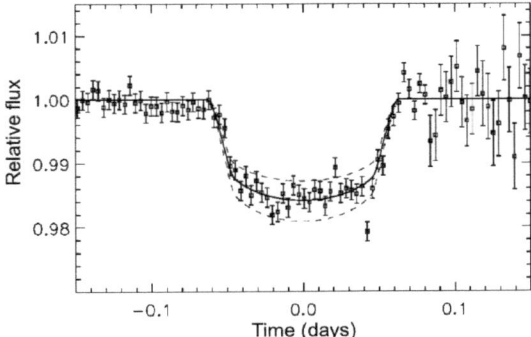

FIGURE 3.2 – The transit of the planet HD 209458 b in front of its star, observed from the Earth. The measurements after the transit are affected by the low altitude of the star above the horizon. The dashed curves would correspond to a radius 10% larger or smaller than the radius determined here, which shows the accuracy of the measurement. The radius of the planet, which is located at 47 pc (143 light-years) of us, is 1.27 ± 0.2 times that of Jupiter and its mass is 0.63 times that of Jupiter. The transit repeats every 3.5 days, which corresponds to the period of revolution of the planet, which is a hot Jupiter. According to Charbonneau, D. et al. (2000).

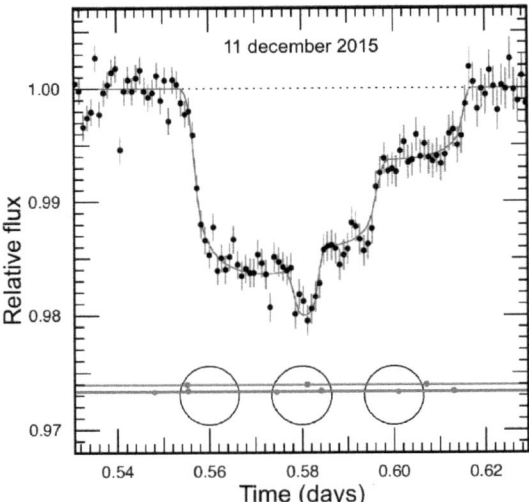

FIGURE 3.3 – Transit of three planets observed with ESO's Very Large Telescope (VLT) in front of the star TRAPPIST-1. The position of the three planets with respect to the stellar disk is indicated at the bottom for three times distant of 0.02 days (28.8 minutes). See Appendix 2 for details on this system, which includes seven near-Earth mass planets. From Gillon, M. et al. (2017) *Nature*, 542, 456.

Observations from the Earth

After the discovery of the transit of HD 209458 b, several programs were set up in order to detect exoplanets directly through transit. These searches are carried out in the framework of systematic campaigns using dedicated telescopes and very stable cameras in order to obtain the required photometric precision. But this research must take a problem into account. The decrease in flux observed in the photometric curves of the stars may be due to causes other than the presence of an exoplanet: in particular, the light curves of the so-called eclipsing binary stars have periodic drops of signal comparable to those induced by planetary transit; these are called a false positive signal. When a candidate exoplanet discovered by transit is suspected, it is necessary to observe its star by velocimetry to confirm the discovery, which is in any case useful to obtain the mass of the planet.

After three new transits were observed at the end of 2003, the number of transits detected increased little by little to reach 27 in 2009. These findings were made by teams of researchers grouped around a dedicated telescope and instrumentation. Several deep surveys have been performed from the ground, including HAT/HATnet (Hungarian Automatic Telescope, some 60 detections in 2017), OGLE in Las Campanas, Chile (initially oriented towards the detection of gravitational lenses), and WASP/SuperWASP (Wide Angle Search for Planets, approximately 140 detections in 2017). Many other programs are underway or under development, including one with the MMT (Multi-Mirror Telescope) in Arizona and SuperLupus at the Siding Springs Observatory in Australia. ESO and the Leiden University are presently conducting a joint program, MASCARA (for Multi-site All Sky CameRA) in which all stars down to magitude 8.4 are observed every 6.4 seconds in the whole sky.

The Antarctic continent is an obviously privileged site because it permits continuous observation of the stars of the southern hemisphere. The A-STEP project was installed at Dome C and has been operational since 2010. It tested the photometric quality of the site and detected several exoplanets candidates; it also highlighted the difficulties inherent in a polar site. Since 2017, it has devoted himself to the study of the β Pictoris system. At Dome A, the highest point in Antarctica, Chinese teams commissioned a set of small telescopes, CSTAR (Chinese Small Telescope ARray) to study stellar variability. At the same time, a more ambitious project including three 50 cm Schmidt telescopes, AST3 (Antarctic Survey Telescopes) has been implemented and is under development, but it has also suffered from site-related weather conditions.

Amateur astronomers are increasingly involved in the search for exoplanets, either in campaigns for observing transits from the ground, or by analyzing data from the Kepler satellite.

The Space Age

Photometric observations from the Earth, even in Antarctica, have an accuracy necessarily limited by the scintillation due to the turbulence of the atmosphere. To escape these limits, it is necessary to observe from space. It is then possible to obtain both long-term continuous observations with a photometric stability reaching an accuracy of 10^{-4} over several weeks.

The CoRoT mission

The first space mission dedicated to the observation of exoplanets was CoRoT (Convection, Rotation and Planetary Transits), a French satellite launched by CNES in cooperation with several partners, mostly European (Figure 3.4). At the origin of the project, in the 1990s, the mission's primary objective was stellar seismology through the study of star variability. However, with the discovery of the first transits, it became apparent that the ultra-precise, long duration stellar photometry method allowed by this satellite was perfectly adapted to the detection of exoplanets.

FIGURE 3.4 – The CoRoT satellite. © CNES.

Launched in December 2006, CoRoT was equipped with a telescope with an effective diameter of 30 cm and a CCD matrix operating in the visible range. It was placed on a polar orbit allowing the observation of two regions of the sky, each for a period of six months. Approximately 12,000 stars were observed simultaneously. After seven years of continuous operation, the mission was interrupted following a breakdown likely caused by a bombardment of high-energy particles. In addition to major results concerning stellar seismology, the CoRoT mission detected some thirty exoplanets in transit in front of their stars.

Due to the observation time of a single field limited to six months and the need to observe two or three successive transits to validate the detection, all exoplanets detected and confirmed have a short period and are thus are located near their star, at a distance of less than 0.3 a.u. But apart from this common characteristic, they present an extreme diversity in terms of orbits, diameters, masses and densities. With a mass of 5 terrestrial masses and a revolution period less than a day, CoRoT-7 b was, at the time of its discovery in 2008, the smallest known exoplanet; its surface day side temperature could exceed 2000 degrees. CoRoT-9 b, with a mass comparable to that of Jupiter and a period of 95 days, was, when discovered in 2010, the first exoplanet presenting an analogy with our cold Jupiter. CoRoT-15 b, which orbits in three days around its star with a mass above 60 times that of Jupiter, is extremely dense, and is more probably a brown dwarf star. If the number of exoplanets detected by CoRoT was modest, the sample of discoveries already revealed the extraordinary diversity of the physical and orbital parameters of these new objects.

The Kepler Mission

In March 2009, NASA launched a space mission dedicated to the search for exoplanets by transit: the Kepler space telescope, of 0.95 m diameter, located in a heliocentric orbit behind the Earth (Figure 3.5). Its field of view, significantly greater than that of CoRoT, allowed the simultaneous observation of 145,000 stars in a fixed field of the Milky Way. In February 2011, the Kepler's data mining consortium announced the detection of 1235 candidate exoplanets. In most cases, the host stars are too weak for a confirmation by velocimetry, but statistical studies of the different classes of exoplanets can already be made. Another important result: Kepler's list also includes many multiple systems, as well as some fifty objects that can be found in the "habitable zone" of their star, where water could be in liquid form (we will return to this in chapter 10). In December 2011, six exoplanets were found around a solar-like star, Kepler-20. Kepler-20 e and Kepler-20 f have a mass comparable to that of the Earth; the others are super-Earths of 10 to 20 Earth masses, or of masses intermediate between that of the Earth and that of Neptune. There is no example of these super-Earths in the Solar System.

FIGURE 3.5 – The Kepler space telescope. © NASA.

For the first time, it appeared that exoplanets of smaller size than Neptune are the most numerous. This result, announced in 2011, was confirmed in the following years: in November 2013, the Kepler's list of candidate exoplanets had about 1400 objects of the size of Neptune, a thousand super-Earths, about 700 exo-Earths and only 200 objects the size of Jupiter. Also in 2013, Kepler announced the discovery of two super-Earths in the habitable zone of a star of solar type surrounded by five planets, Kepler-62.

After CoRoT, Kepler has shown systems of extreme variety, including planets orbiting a double stellar system, either outside the system, or orbiting one of the stars of the couple. In April 2017, 2327 exoplanets out of a total of 3610 had been discovered by Kepler and received confirmation. The distribution of these planets according to their volume (the least biased of the quantities that can lead to a statistic is the radius of the planet), brings important surprises: roughly 50% are super-Earths whose volume is between three and 30 times that of the Earth. The larger planets constitute only 30% of the total, of which only 12% have a larger volume than our Jupiter, whereas the first discoveries suggested that these large planets were the most frequent. The proportion of terrestrial planets smaller than three times that of the Earth (30%) is undoubtedly underestimated, as they are the most difficult to detect. It must be borne in mind, however, that all these planets have periods of revolution of less than a few years, for at least three transits must be observed in order to be sure of the result: we have, therefore, only a partial view of the panorama of exoplanets.

This avalanche of results also makes it possible to specify the probability for a star to host at least one planet. It appears today that for the stars of our Galaxy, this probability is larger than 50%. More should be learned after the GAIA space mission, which has significant photometric capabilities and is expected to detect thousands of exoplanets by transit, in addition to the even more numerous ones that will be detected by astrometry.

For some of these exoplanets, the transits are not quite regular, implying the presence of another perturbing planet. Seven new exoplanets were discovered by this method, called TTV, for Transit Timing Variation.

Primary and Secondary Transits

In the preceding section, we have called "transit" the passage of an exoplanet in front of its star. This is in fact the optimal configuration for detecting its presence, because it results in a decrease of the flux of the star of the order of 1%, measurable from Earth in the case of giant exoplanets. However, another phenomenon also deserves attention: it is the passage of the planet behind its star. This also results in a decrease in the flux observed, by an amount equal to the flux emitted by the planet itself. In the first case (passage in front of the star), one speaks of "primary transit" or occultation (of the light of the star); in the second case (passage behind the star), we speak of "secondary transit" or eclipse.

Let us consider what happens to the exoplanet along its orbit during an entire revolution (Figure 3.6). When it comes in front of its star (primary transit), it presents its night side to the observer. When it emerges and moves away towards its maximum elongation, it presents to the observer an illuminated quarter that gradually increases until reaching half of its disk. When it again approaches its star to pass behind it (secondary transit), the illuminated part increases until it fills the entire disk. Just before and after the eclipse, the exoplanet presents to us its full day side. The information that can be extracted on the nature of the star is therefore very different in both cases. We have seen that the primary transit informs us about the ratio of the radii of the exoplanet and its star. In the case of secondary transit, we have a measure of the ratio of the fluxes emitted in the visible or the infrared by the exoplanet and its star.

In the early 2000s, astronomers focused on measuring the integrated flux in the visible domain, first in primary transit with the objective of detecting new exoplanets and then in secondary transit to obtain additional information on the temperature of the exoplanet. A few years later, it appeared that the spectrum (that is to say the measurement of the radiation as a function of the wavelength) of the flux difference observed during the primary and secondary transits could provide valuable information on the nature of the atmosphere of exoplanet. Thus, a field of research that is currently developing was born: the spectroscopy of exoplanets through transit.

The method of transits

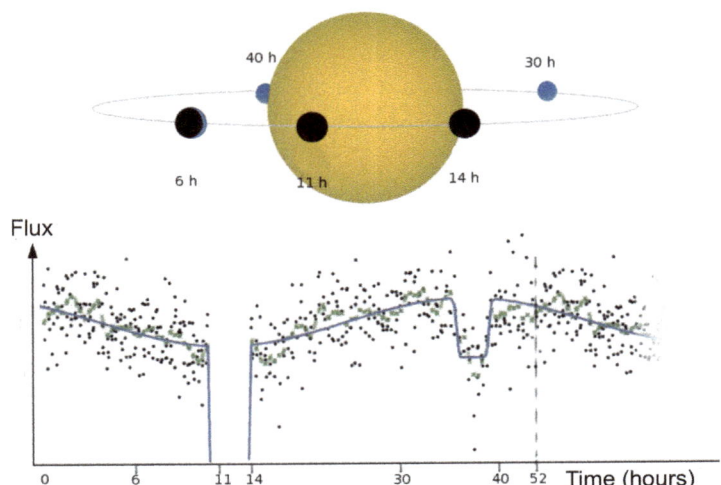

FIGURE 3.6 – Phase curve of a star in the case of a primary transit (11–14 h) and then a secondary transit (35–40 h). The gradual variations in the luminosity between the transits are due to the fact that we see a variable fraction of the illuminated part of the planet. Diagram of the authors.

Transmission spectroscopy (primary transit)

When an exoplanet with an atmosphere passes in front of its star, the light of this star is attenuated not only by the disk of the planet but also by a gaseous ring surrounding the planet, which absorbs more or less at the different wavelengths according to the composition of the atmosphere. The fraction of the light absorbed by the exoplanet and its atmosphere can be measured by difference of the fluxes observed during transit and out of transit, just before or just after the transit. Its variation with wavelength, e.g. its spectrum, gives us interesting information about the nature of the atmosphere.

The amplitude of the signal depends on the width of the ring corresponding to the atmosphere of the exoplanet. It can be shown that this width is proportional to the scale height H of the atmosphere, which is defined as the altitude z at which the atmospheric pressure P decreases by a factor e ($P = P_0\, e^{-z/H}$). The scale height is proportional to the temperature and inversely proportional to the average molecular mass of the gas and to the gravity of the exoplanet. The "hot Jupiters", predominantly made up of hydrogen and helium of low molecular mass, have thick atmospheres and are the best candidates for transmission spectroscopy. Note also that the ring observed during a primary transit informs us about the state of the atmosphere at the terminator, that is to say at the border between day and night on the exoplanet. This fact is especially interesting in the case of hot Jupiters, because these objects, very close to their stars and most often in circular orbit, have the particularity of always presenting the same face to their star, as does the Moon with respect to the Earth. A very strong temperature contrast between the day and night sides must ensue, and it is difficult to imagine what is going on at the terminator, the region that is being explored during a primary transit.

For a warm Jupiter, the amplitude of the measured signal is of the order of 10^{-4} for the brightest stars. To measure such a signal with sufficient accuracy, it is necessary to go into space. Transmission spectroscopy measurements were made possible by two space observatories, the Hubble Space Telescope (HST) and the American Spitzer satellite. The measurements were carried out in the visible range and near infrared with the HST and in the near and middle infrared, up to 20 micrometers, with Spitzer. Neither HST nor Spitzer had been designed for measurements of such accuracy, and the development of specific algorithms was necessary to extract the tiny signal from the different sources of noise. The first results were obtained for two giant exoplanets whose star is particularly bright, HD 209458 b (we have already met it, this is the first exoplanet observed by transit) and HD 189733 b (Fig. 3.7), another giant exoplanet with a mass of 1.1 $M_{Jupiter}$ and an orbital period of 2.2 days. In the first case, atomic hydrogen was detected, indicating the presence of a strong atmospheric escape, atomic sodium, and several molecules (methane CH_4, carbon dioxide CO_2, water H_2O). In the second case, water vapor, carbon monoxide CO and methane were measured. These results are presented in more detail in Chapter 8.

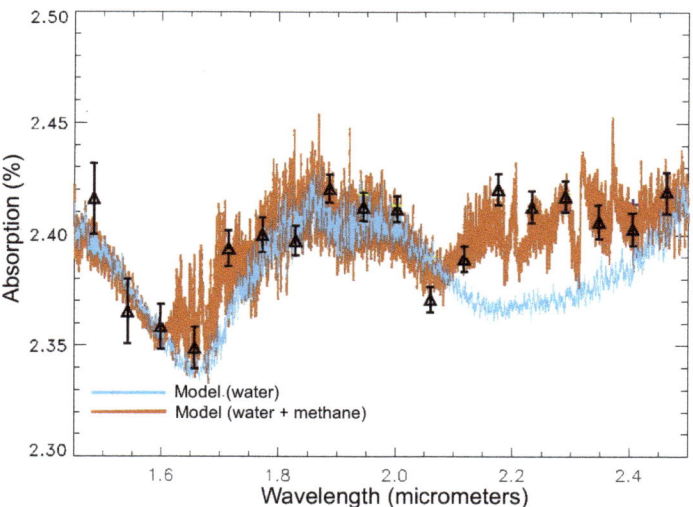

FIGURE 3.7 – Evidence for water vapor and methane in the atmosphere of hot Jupiter HD 189733 b by transmission spectroscopy. The black triangles with error bars represent the measurement points. The curves correspond to atmospheric models containing H_2O (blue) and $H_2O + CH_4$ (orange). From Swain, M.R. et al. (2008).

Emission spectroscopy (secondary transit)

The emission spectroscopy is very interesting because it is done in the infrared, where the measured signal informs us about the temperature of the exoplanet. The measured ratio F_P/F_* of the fluxes of the planet F_P and of the star F_* is related to the ratio of the temperature T_P of the planet and of the temperature T_* of its star. In the visible and near infrared range, the ratio F_P/F_* is approximately equal to $(r_P/r_*)^2(T_P/T_*)^4$, according to Stefan's law, r_P and r_* being respectively the radius of the planet and of the star. In the mid infrared (about 20 micrometers and beyond), the ratio F_P/F_* is approximately equal to $(r_P/r_*)^2(T_P/T_*)$ according to the Rayleigh-Jeans law. In the first case, the expected ratio for a hot Jupiter is about 10^{-5}. It is about ten times stronger in the second case because the contrast between the temperature of the planet and that of the star increases substantially with the wavelength. An example of emission spectrum is presented in Figure 3.8.

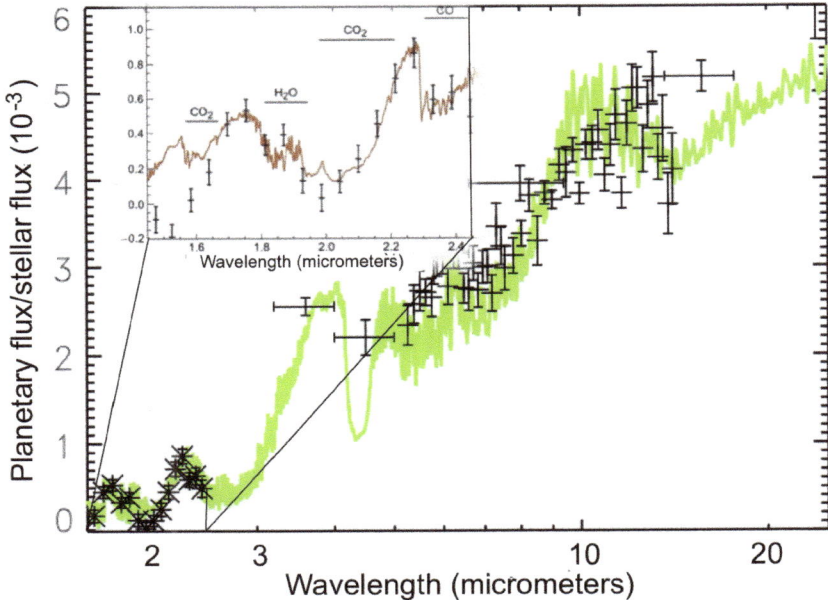

FIGURE 3.8 – Example of emission spectrum obtained on the exoplanet HD 189733 b. The red or green curves correspond to the predictions of an atmospheric model, and the points with error bars to the observations. The smaller rectangle contains an enlargement, in linear wavelength scale, of the near-infrared data obtained with the Hubble Space Telescope. The data in the mid infrared come from the Spitzer satellite. From Tinetti, G. & Griffith, C. (2010), in *Pathways Toward Habitable Planets*, Coudé du Foresto, V. Gelino, D., Ribas, I., eds., *ASP Conference Series*, Vol. 430, 115.

In the infrared domain, the identification of atmospheric constituents requires the use of radiative transfer models and the simultaneous determination of the thermal profile of the atmosphere, which plays a dominant role in the shape of the spectrum. The molecular signatures may appear in emission or in absorption according to the temperature gradient of the atmospheric region in question. The results obtained by emission spectroscopy are detailed in Chapter 8.

After the spectacular success of the Kepler mission, the search for exoplanets by transit is oriented towards objects around stars that are bright enough to allow the spectroscopic observation of planetary transits. The first of these missions is the TESS (Transiting Exoplanet Survey Satellite) mission, launched in April 2018, and also CHEOPS (CHaracterizing Exoplanet Satellite), a small mission led by ESA in partnership with Switzerland and presently in operation. Then there will be more ambitious missions allowing spectroscopic observations by transit. The JWST (James Webb Space Telescope), the successor to the HST, is expected to make a significant contribution to this research. Projects for telescopes dedicated to the spectroscopy of exoplanets by transit are also under study. In particular, the ARIEL (Atmospheric Remote Sensing Infrared Large Survey) project has been selected under the Cosmic Vision program. The launch of the mission is planned for 2028. These projects are presented in more detail in Chapter 11.

Gravitational transits

In parallel with the observations of velocimetry and transit, a third method of detecting exoplanets has emerged in the first decade of the 21st century: observation of gravitational transits, or gravitational microlenses. What is it about? The technique is based on an application of Einstein's theory of general relativity, which predicts the curvature of light rays as they pass close to a massive object. The gravity field of the object can be compared to a lens that curves the path of nearby light rays, hence the name "gravitational lens" given to this phenomenon. If a star passes exactly in front of a distant star, the light of the latter is amplified at the time of passage, the star acting as a lens converging the light rays from the distant star towards the observer.

The first application of microlensing was the systematic search for very low mass stars of our Galaxy, the famous brown dwarfs, which were suspected to be responsible for the missing mass, the material whose presence is revealed by the dynamics of our Galaxy but which escapes detection. The program involved a deep systematic survey of the sky in the regions richest in stars, like the center of our Galaxy. Since the early 1980s, this research has led to the detection of a large number of gravitational transits, but these were due to normal stars: the search for brown dwarfs by gravitational

The method of transits

transit was negative. Thus, the brown dwarfs are not responsible for the missing mass revealed by the dynamics of our galaxy.

The search for exoplanets by gravitational transit is derived from this method. If the star that plays the role of a lens possesses an exoplanet that also passes in front of the distant star, the system behaves like a double lens and the light curve of the distant object exhibits, in addition to the amplification due to the star, a narrow peak due to the exoplanet (Figure 3.9). This phenomenon was observed for the first time in 2004 around a star located at 5000 parsecs, allowing the identification of an exoplanet of mass equal to 1.5 times that of Jupiter, at 3 a.u. of its star. A year later, another object was discovered by the same technique; later observations of the lensing star by the HST led to the specification of the exoplanet, with a mass of 4 $M_{Jupiter}$, at a projected distance of 3.6 u.a. from its star. In 2006 came a new surprise: a French team announced the discovery in 2005 of an exoplanet of only 5.5 Earth masses, located at 2.6 a.u. of its star; it was the smallest exoplanet known at the time. The detections have continued in the following years.

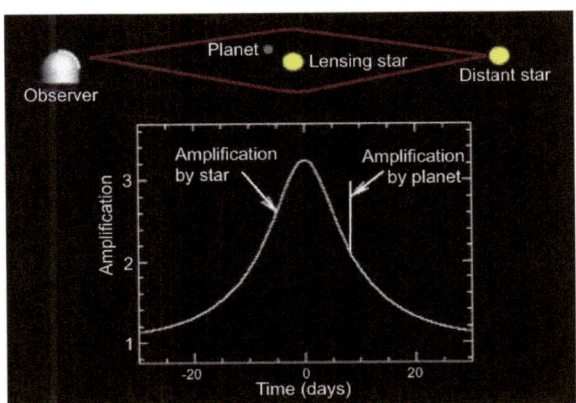

FIGURE 3.9 – Schematic diagram of the detection of an exoplanet by gravitational amplification. At the top, deflection of the light rays when the lens star is exactly aligned with the source star (the deflection by the planet is not represented). At the bottom, the star-lens light curve showing the peak associated with the presence of an exoplanet, when the star-planet set moves downward from the source star.

In 2020, some 120 exoplanets, including five multiple systems, have been discovered by the technique of gravitational transit. Most of the objects have been discovered by the OGLE program devoted to the search for microlenses. It is in fact the same technique – the deep photometric study of a stellar field – that is used for normal transits and gravitational transits. The difference is that the presence of an exoplanet results in a periodic absorption in the case of planetary transits and gives a secondary peak in the light curve in the case of gravitational transit. This second signature is not repetitive because once the star responsible for the amplification has passed in front of the background source, there is no possibility of re-observing the phenomenon; this is the main limitation of the method of gravitational transits. On the other hand, it has the advantage of being sensitive to objects of low mass (as shown by the detection of 2005), and of being able to detect exoplanets located at great distances from us, to several thousand parsecs. The sampling carried out by the technique of gravitational transits is therefore very different – and perfectly complementary – from those based on velocimetry or planetary transits.

In the future, the exploration of exoplanets by gravitational transits should acquire a new dimension thanks to the coupling of this objective with the observation of gravitational lensing of distant galaxies located behind large masses, such as clusters of galaxies. Both investigations require the same instrumentation, for a deep photometric probing of a large field in the visible and infrared. Two space projects are under preparation for this program. In the United States, the WFIRST (Wide Field Infrared Survey Telescope recently renamed Roman) mission, which is currently under study, could be launched in the mid-2020s. On the European side, the Euclid mission is developed since 2011 by the European Space Agency within the framework of its scientific program "Cosmic Vision" for a launch planned for 2022.

Bibliography

Charbonneau, D. et al. (2000) Detection of planetary transits across a Sun-like star, *Astrophysical Journal* 529, L45, http://cdsads.u-strasbg.fr/abs/2000ApJ...529L..45C

Bond et al. (2004) OGLE 2003-BLG-235/MOA 2003-BLG-53: a Planetary Microlensing Event, *Astrophysical Journal* 606, L155 http://cdsads.u-strasbg.fr/abs/2004ApJ ... 606L.155B

Chapter 4
Detecting and viewing exoplanets

At the time of writing, about 3700 exoplanets have been discovered. Sixty percent appear, for now, alone around their star, the others are part of multiple systems. Three-quarters were detected by their transit in front of the star, 20% by velocimetry, and the remaining 5% by direct observation or by gravitational disturbances to other planets. But, of course, many exoplanets have been observed by several methods at the same time, which makes their detection more secure and provides very interesting information, as these methods are complementary. Each has its limits and its biases, which are interesting to discuss.

Advantages and limitations of indirect methods of detection

The transit method

Let us recall that this method has the advantage of giving not only the period of revolution of the planet (provided that at least two successive transits have been observed), but also the radius of the planet if that of the star is known from its spectral classification, which is generally the case. If the mass of the star can also be approximately determined from this classification, which is also most often the case, we can calculate, from the period of revolution, the semi-major axis of the orbit of the planet using Kepler's third law (see Box 2 in Chapter 2). But the transit method does not provide the mass of the planet.

A transit is easier to detect if the planet is bigger, and the star smaller. This produces biases in the detection: many planets similar to the Earth or smaller are missing. Moreover, it is necessary to observe several transits to confirm the detection, and as far as possible the secondary transits, when the planet passes behind the star. Indeed, all the stars are more or less variable and we cannot be sure that a temporary decrease of their flux is not intrinsic to the star. This introduces another bias: we miss the planets with long periods of revolution. For example, the Kepler satellite, which enabled exoplanets to be discovered systematically by transit in a certain area of the sky, was launched in March 2009 and ceased to observe the area in May 2013. Certainly, this produces a bias against the detection of planets whose period is greater than one year. It has been possible to confirm a large number of Kepler detections by observing the star by velocimetry, but even in this case there is a bias against long periods because it is necessary to wait until the radial velocity of the star has varied sufficiently to be certain of the detection.

Finally, when we check by velocimetry the validity of detections of exoplanets by transit, we find that quite a large number of these detections are false or correspond to other types of objects than planets, for example binary stars where one of the components has a low mass, or eclipsing binaries. A recent study shows that, despite all the precautions taken in analyzing the observations with Kepler, more than half of the detections verified by velocimetry do not correspond to exoplanets. We must then remain very cautious in the conclusions that can be deduced from these detections. However, the same study provides a complete sample of giant exoplanets, which unfortunately includes fewer than 100 objects. We will use it to specify the properties of these planets.

The probability to observe a transit is obviously proportional to the radius R_* of the host star. As the orientation of the orbits of the exoplanets is random, the probability that an exoplanet passes in front of its star is also proportional to the inverse of the semi-major axis a of its orbit. Thus, in total, the probability of detection by transit is proportional to the ratio R_*/a. We can correct for this bias and find the real distribution of exoplanets according to their period of revolution, as will be done in the next chapter for giant planets.

Exoplanets detected as gravitational microlenses

The microlensing method, which has the advantage of providing the mass of the exoplanet, also presents significant biases. It is easier to detect large-mass planets: with a few exceptions, the mass of the exoplanets detected in this way is far larger than that of the Earth. On the other hand, the amplification they produce on the light of the distant star depends on their position on the line of sight between the Earth and this star. An enormous disadvantage: the observation is unique, for it is not possible at present to find the planet and often even its star after its passage. Thus, the nature of the host star is rarely known, and the period of revolution of the planet and the dimensions of its orbit remain totally unknown.

Exoplanets detected by velocimetry or astrometry

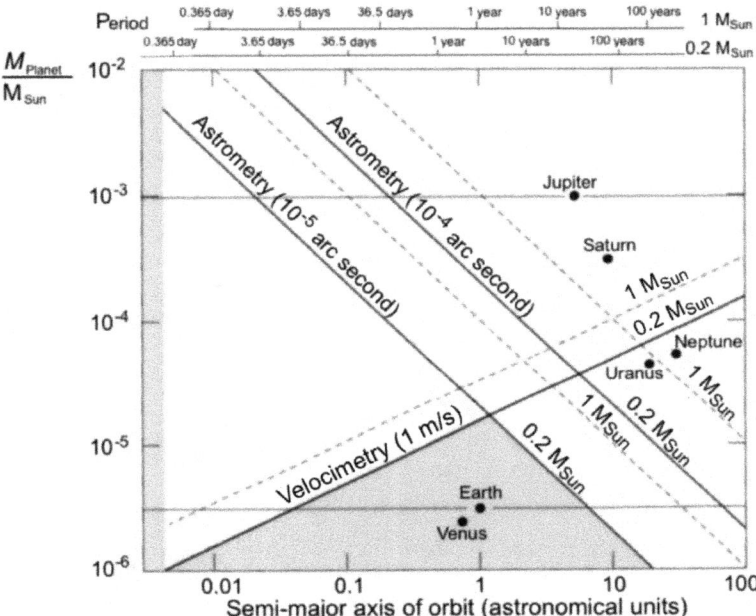

FIGURE 4.1 – The possibilities of detection of exoplanets by dynamic methods. The mass of the planet is plotted as a function of the semi-major axis of its orbit, in logarithmic scales. Above, the period of revolution as a function of the semi-major axis, around a star of 1 and 0.2 solar mass. The lines ascending to the right, corresponding to the velocimetric method, are plotted for a half-amplitude of the radial velocity of 1 m/s for stars of 1 and 0.2 solar mass. This corresponds roughly to the current limits of detection. In this case, the distance does not intervene. The descending lines, for the astrometric method, indicate for a distance of the star of 10 parsecs and stellar masses of 1 and 0.2 solar mass, a half-amplitude of the variation of respectively 10^{-4} and 10^{-5} arc second. The first value can be reached in observations from the ground, the second by the GAIA satellite. The planets of the Solar System are shown. We see that the Earth and Venus would be undetectable at 10 pc distance, but an Earth with an orbit of radius 1 a.u. around one of the nearest stars would be marginally detectable. Authors' schema, inspired by the TOPS report.

Figure 4.1 illustrates the capabilities of these two methods for the detection of exoplanets. They are very different. The detection by velocimetry is easier if the period T, and therefore the radius of the orbit of the exoplanet, is smaller, since the orbital velocity is then larger: it varies as $T^{-1/3}$. On the other hand, astrometry works better as the period is longer, the lateral orbital displacement of the star varying as $T^{2/3}$. Velocimetry does not depend on the distance of the star (however, the signal is weaker if the star is more distant). On the other hand, astrometry depends directly on the distance, since it is an angle that is measured: the closer the star is, the more sensitive this method is. In both cases, there is a bias in favor of short-period exoplanets because they require less prolonged observations.

While no planet has yet been detected by astrometry, it has a bright future, as recent advances are dazzling. The GAIA astrometric satellite already makes it possible to achieve an accuracy of 10^{-5} arc second on the position of the stars brighter than magnitude 12. It should detect many planets with a mass greater than a dozen earth masses – super-Earths, Uranus, Saturns and Jupiters, provided that their period is not much longer than the current satellite lifetime of 5 years. This sample will be usable for statistical purposes. Astrometry by imaging from the ground, which measures the position of a star with respect to its immediate neighbors, achieves an accuracy of 10^{-4} arc second. Visible or infrared interferometry with a large base on the ground should make it possible to do better, but at the cost of a great complexity of measurement. In both cases, patience will be required, as the measurements will have to be repeated many times over the years.

Direct observation: a very difficult problem

All the techniques that we have just seen are indirect, since the presence of a planet is detected by its effects. It would be far more satisfactory, and more effective in understanding the nature of exoplanets, to detect them directly, that is, by receiving and analyzing the light they emit. But in order to see a planet that gravitates around a star, we have to overcome two major obstacles. The first is that the star is enormously brighter than the planet, by a factor of the order of a billion in visible light. The second is that, except for rare exceptions, the planet is very close to the star, so that the spreading of images by the turbulence of the Earth's atmosphere does not allow it to be separated from its star if this turbulence is not corrected or suppressed.

The current techniques are fortunately able to overcome at least some of these obstacles. To overcome the first one requires to hide as well as possible the image of the star, without occulting that of the planet. There are two possibilities for this that we will describe below: coronography and black-fringe interferometry. To overcome the second obstacle requires either to go to space or, for ground observation, to correct in real time the effects of atmospheric turbulence by adaptive optics. The ideal, realized in the most recent instruments like SPHERE at the European Southern Observatory, is to combine several techniques, in this case coronography and adaptive optics.

It is interesting, and often necessary, to observe in the infrared. In fact, it is easier to work with a coronograph or an interferometer if the wavelength is larger, and adaptive optics is then much more efficient and actually allows

Detecting and viewing exoplanets

to reach the theoretical resolving power of the telescope. And, above all, as the example of Figure 4.2 shows, the contrast between the planet and the star improves in the infrared: if the planet is hot enough, its thermal radiation in the middle infrared and even in the near infrared facilitates considerably its detection. If the planet is young and still in the phase of contraction, it can emit by itself in the infrared and would be easy to observe. But if it is old, it emits only by diffusion of the radiation of the star, and can be detected only if it is close to it.

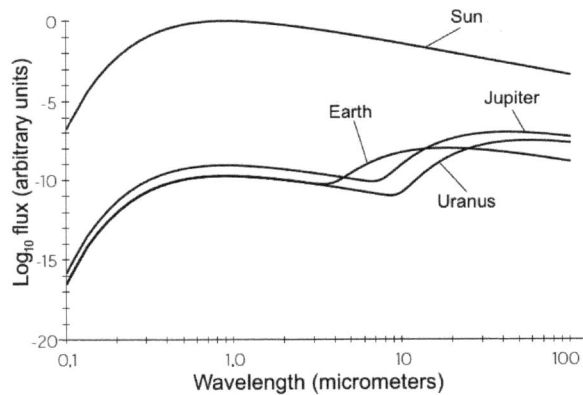

FIGURE 4.2 – Distribution of spectral energy of the Sun, Earth, Uranus and Jupiter, supposedly seen from a large distance. The ratio of the planetary flux to the solar flux, which is very small at wavelengths where the planet only diffuses the solar radiation, increases considerably at larger wavelengths where its thermal emission comes into play. If the planet is hot by itself, it is easier to detect. From the TOPS report.

Coronography

Coronography is a technique invented around 1930 by the French astronomer Bernard Lyot (1897–1952) in order to observe the solar corona outside the eclipses, by hiding as well as possible the solar disk, which is much brighter than the corona. It is not sufficient to place a mask on the image of the disk, because this does not eliminate the light scattered by the edges of the objective of the telescope and the defects of the glass, as well as a central spot due to the reflection between the two faces of this lens. This last defect disappears if a mirror telescope is used, but others are introduced, as there is light diffracted by the supports of the secondary mirror. The trick of Lyot was to place a mask on an image of the objective, that is to say a pupil in optician's language (Figure 4.3).

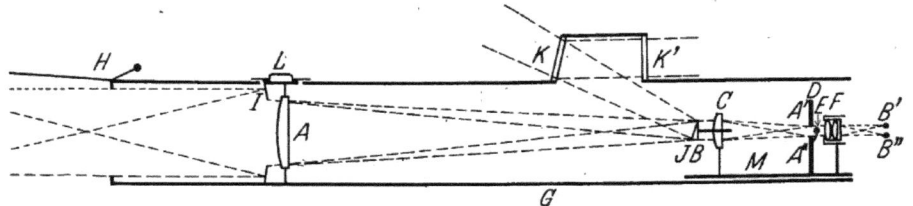

FIGURE 4.3 – Diagram of Lyot's coronograph. The light of the Sun falls on the objective A, a single lens which makes its image in B. A mask is placed on this image, and the light is sent to the outside in K by the mirror J fixed on the mask. The field lens C makes the image of the objective in $A'A''$, where the scattered light is stopped by the diaphragm D. A small central screen E cuts off the light produced by reflection on the faces of the objective A (if the objective is the mirror of a telescope, a suitable diaphragm stops the light diffracted by the supports of the eventual secondary mirror). The lenses F form the image of the corona in $B'B''$. From Lyot, B. (1932).

This assembly is nevertheless insufficient for our aim, and several improvements are needed. One of them consists of decreasing the transmission of the lens towards the edges, which degrades a bit the angular resolution but eliminates better the light diffracted by the edges of the objective. It is also possible to place on the pupil a phase mask formed by four contiguous squares: two of the opposite squares have a thickness slightly different from the other two so as to introduce a half-wavelength difference in optical path: this produces a destructive interference to the image of the star. By these methods, the image of the star can be attenuated by a factor of 90,000 without affecting its surroundings; by putting three phase masks in series, this factor can reach $1.5 \ 10^8$.

The black-fringe interferometer

Another way of reducing the light of the star, which has some relation to the phase mask that we just described, is to use a two-telescope interferometer (Figure 4.4). The recombination of the beams produces an image modulated by interference fringes, as in the famous experiment of Young's slits (Figure 4.5). By adjusting the phase difference between these beams so that the image of the star is on a black fringe, it is made to disappear. It is now necessary to be able to rotate the interferometer and change the spacing between the two telescopes so that the image of the planet is on a bright fringe. Difficulties arise from the sensitivity of the phase difference to wavelength, and also from the apparent diameter of the star if it is resolved. The use of fringes given by several telescopes makes it possible to overcome them, at least in part. The DARWIN space interferometer project used this principle with four telescopes but has been abandoned due to its complexity.

Detecting and viewing exoplanets

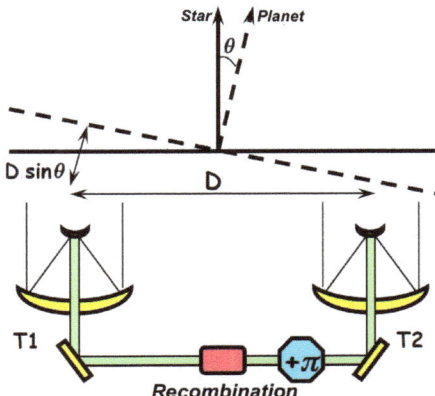

FIGURE 4.4 – Interferometer with central black fringe. The star is pointed to the two telescopes whose light is sent to a central point. The recombination produces interference fringes that modulate the image of the observed region. If the optical paths are equal, the star is on a white fringe. A phase difference of π is introduced into one of the beams so that the star is now on a black fringe and disappears, which makes it possible to observe much weaker objects such as exoplanets in the bright fringes: see the next figure. From Daniel Rouan, with thanks.

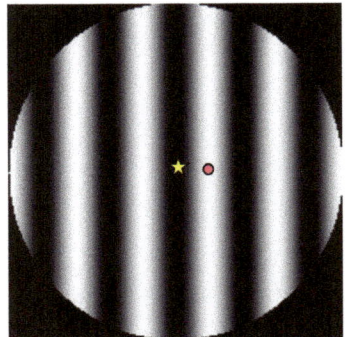

FIGURE 4.5 – Recombination image given by the interferometer of the previous figure. The star is on the central black fringe and is not visible, while a planet (red) is on a bright fringe. From Daniel Rouan, with thanks.

How to obtain perfect images: adaptive optics

If the absence of atmosphere makes it possible to obtain perfect images from space, limited only by the diffraction of the objective, the situation differs for observations from the ground: the turbulence of the atmosphere spreads out the images and makes them unstable, so that it is rarely possible to see the diffraction rings around the image of a star; this is all the more difficult if the instrument has a large diameter. However, a technique allows to correct in real time the effect of disturbances of the atmosphere and thus to obtain almost perfect images: adaptive optics. It is particularly useful for the detection of exoplanets which are faint point objects in the immediate vicinity of a much brighter point object, the host star of the planet.

The principle of adaptive optics is simple, but its realization is rather heavy. At least every one-hundredth of a second, the deformations of the wave front entering the telescope from the star are analyzed and compensated by deforming a thin mirror placed in the path of the light. The calculation and deformation of the mirror must be done very rapidly so that the wave front has not appreciably changed between measurement and compensation; this requires a powerful computer. Figure 4.6 illustrates this principle. If the realization is done correctly, the images can be perfect, that is to say limited by diffraction, in a field of a few arc seconds of degree around the star, which is sufficient for the detection of exoplanets. Adaptive optics works very well in the near infrared, but poorly in the visible, so that the observations are generally made in the infrared, which is not a disadvantage for the detection of exoplanets. Figure 4.7 shows the first image of an exoplanet obtained at the European Southern Observatory using the NACO adaptive optics system.

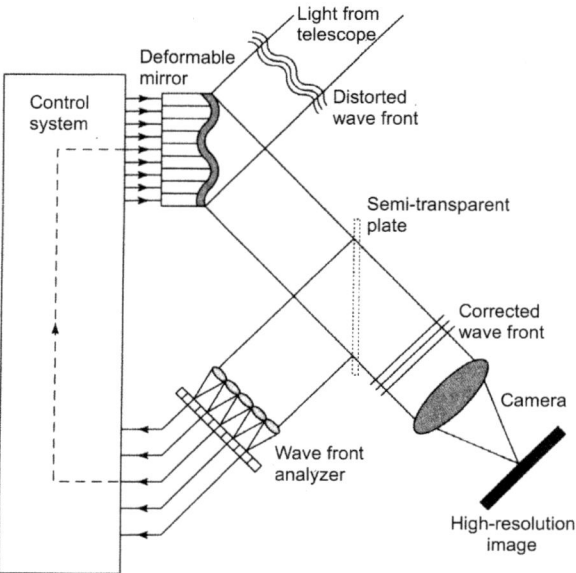

FIGURE 4.6 – Principle of adaptive optics. Explanation in the text. © Paris Observatory.

Detecting and viewing exoplanets

FIGURE 4.7 – The first detection of an exoplanet by direct imaging in the infrared, carried out with the NACO instrument of the ESO's Very Large Telescope in Chile, with adaptive optics but without coronography. The star, AB Pictoris, is a brown dwarf of 0.025 solar mass, very faint, which considerably facilitated the detection of the planet. This planet has a mass equal to 5 times that of Jupiter and is located at 0.78 arc seconds from the star, corresponding to 55 astronomical units. ESO, from Chauvin, G. et al. (2005).

Combined coronography and adaptive optics

At present, the direct observations of exoplanets always combine these two techniques. A remarkable example (Figure 4.8) is provided by the direct infrared imaging of the exoplanet that orbits around the β Pictoris star, famous for its debris disc. The images taken a few years apart show the orbital motion of the planet.

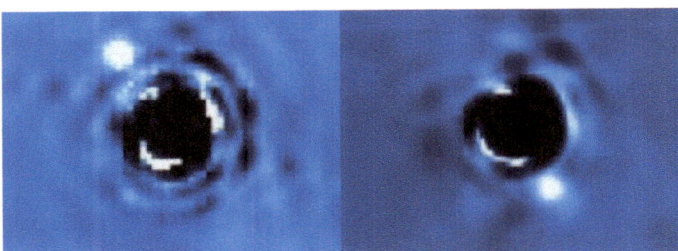

FIGURE 4.8 – The planet around β Pictoris, observed with the NACO instrument of the VLT in two observations separated by several years: October 2003 (top) and November 2009 (bottom). The distance from the planet to the very bright central star (here hidden by a coronographic mask) is in both cases about 8 a.u., or 0.40 arc seconds (at the distance of the star, 19.3 parsecs). Between the two observations, the planet has traveled about a third of its orbit, which is quite eccentric. Its mass is about 7 times that of Jupiter. From Lagrange, A.-M. et al. (2009, 2010).

An instrument specialized in planetary imaging, known as SPHERE (for Spectro Polarimetric High-contrast Exoplanet REsearch), was built by the European Southern Observatory and has been operating since 2014 on one of the four 8-m diameter telescopes of the VLT (Very Large Telescope). It includes a high-performance adaptive optics system with a correction rate for atmospheric perturbations of about 1 millisecond and contains the different types of coronographs that we mentioned above. It works in the near infrared in direct imaging, but it also has polarimetric capabilities. Indeed, the light reflected by the planets is polarized while that of the stars is not, which can help in the detection. It also includes narrow spectral band imaging and spectroscopy capabilities to determine the properties of exoplanets and their atmosphere. Figure 4.9 shows the first exoplanet discovered with this instrument.

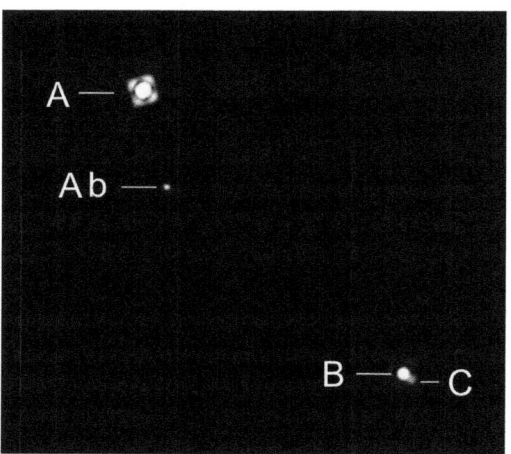

FIGURE 4.9 – The first exoplanet discovered with ESO's SPHERE instrument in 2015 in the triple stellar system HD 131399 A, B and C. The dimensions of the image are 5″ × 5″, that is to say 490 × 490 a.u. This infrared image has been obtained in two steps: one for all the stars, the other for HD 131399 A and the planet HD 131399 A b, which is actually 200,000 times fainter than this star. It has a mass about 4 times that of Jupiter and its temperature is 850 K (577 °C), which makes it a relatively warm giant planet, presently undergoing cooling. The planetary orbits in such a configuration are considered to be generally unstable. The age of the system is estimated as 16 million years. ESO, from Wagner, K. et al. (2016).

Instruments similar to SPHERE are mounted on the American Gemini telescopes and on the Japanese telescope Subaru. They all allow to obtain a low-resolution spectrum of the exoplanets they observe. Moreover, the position of these planets can be measured with the accuracy of astrometry on the ground, which can reach 10^{-4} arc second as we have seen. Repeated observations are beginning to allow the determination of the orbit of such exoplanets, as is the case with β Pictoris b and the four planets surrounding HR 8799 (see Appendix 2). As we write, some 30 planets have been discovered by imagery, and images exist for 89 exoplanets. Note that the term "image" may be misleading, for in these images of planetary systems the planets cover no more than one pixel...

A new track for the future: the search for exoplanets in the radio domain

There is a spectral domain in which the ratio of the flux of the planet to that of the star is particularly favorable: it is that of the decametric emissions which, in the case of Jupiter, are comparable to those of the Sun itself. These short-term emissions (less than 300 milliseconds), linked to the presence of a planetary magnetic field, are in the low-frequency radio domain (less than 100 MHz). They have been sought for more than ten years, especially with the UTR-2 telescope of Kharkiv in Ukraine, around hot Jupiters, these planets being supposed particularly favorable to this type of emission.

To date, this search has not produced any detection. However, the situation could change with the forthcoming Square Kilometer Array (SKA), which is expected to be operational in the 2020s. This instrument consists of two giant networks of antennas, whose construction is in progress in South Africa (for frequencies 50–350 MHz) and Australia (350 MHz–25 GHz). The sensitivity of SKA will be such that the equivalent of our Jupiter located at 10 parsecs should be detectable in one hour of integration time. The advantage of this new type of detection is that it would demonstrate the existence of a magnetosphere around the exoplanet, which is not without consequence for exobiology, because it is commonly accepted that magnetospheres protect the atmospheres from the stellar flux of high-energy particles, making a priori the environment more favorable to the appearance and development of life.

Bibliography

Beuzit, J.-L. et al. (2006) SPHERE, a planet-finder instrument for the VLT, *The Messenger* 125, 29, https://www.eso.org/sci/publications/messenger/toc.html?v=125&m=Sep&y=06

Sahlmann, J. et al. (2014) Astrometric planet search around southern ultra-cool dwarfs I., *Astronomy & Astrophysics* 565, A20

Perryman, M. et al. (2014) Astrometric exoplanet detection with GAIA, *Astrophysical Journal* 797, 14

Coughlin, J.L. et al. (2016) Planetary Candidates Observed by Kepler. VII. The First Fully Uniform Catalog Based on the Entire 48-month Data Set (Q1-Q17 DR24), *Astrophysical Journal Supplement* 224, 12C

Chapter 5
The variety of exoplanets

Before examining in detail the different classes of exoplanets, we will cite a number of highlights that have emerged over the years, each of which is a revolution in itself. These results relate to the frequency of exoplanets, their distance to their star, the variety of their orbits, their presence in multiple systems, and even their presence around or within double stellar systems.

The outstanding results of the last twenty years

A multitude of exoplanets

Statistics on the number of stars with planets were made from Kepler satellite data, from velocimetric studies, and from gravitational microlensing measurements. The probability for a solar-like star to have one planet or more is at least 50%, and it could even reach 100% for the M stars, which make up 90% of the entire stellar population. Any star might therefore be accompanied by a planet! Obviously, each method has a bias and concerns only a particular sample. The velocimetry and transit methods are concerned with stars relatively close to us, whereas gravitational microlensing detections, which concerns objects at a great distance in the halo or bulb of our Galaxy, have been made only for a small number of objects. Nevertheless, all indications are in the same direction: the formation of planets seems extremely common – as common in fact as the formation of a protoplanetary disk – in the life of a star.

Giant exoplanets very close to their stars

We have mentioned above (Chapter 2) the enormous surprise that was the discovery of the first exoplanet around a star of solar type, 51 Peg b: this giant exoplanet revolves around its star in just four days! The whole scenario of the formation of planets, built to account for the structure of the Solar System, was questioned. The successive discoveries that followed that of 51 Peg b confirmed the existence of this new class of objects. Twenty years later, we have a good statistics on the orbital distance or period of these giant exoplanets; several hundred objects belong to this new class, called "Hot Jupiters" (Figure 5.1). We will see later (Chapter 7) how migration within the planetary disk makes it possible to explain the existence of this class of objects.

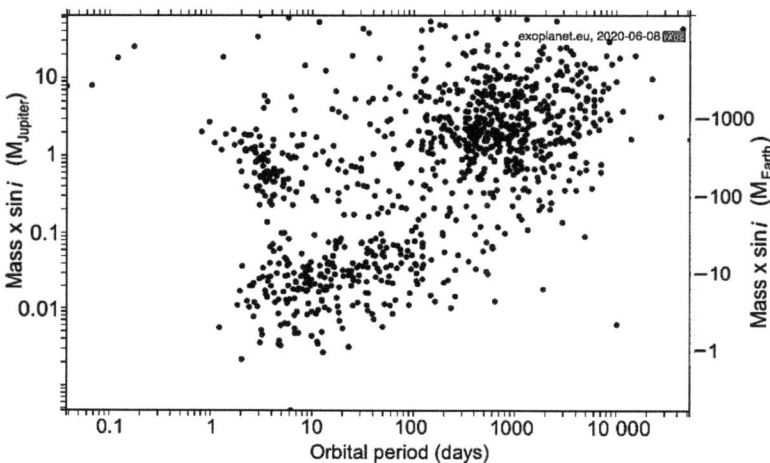

FIGURE 5.1 – Mass, or product of the mass by $\sin i$, as a function of the period of revolution of exoplanets of known mass. This figure should be considered as only indicative, because it does not correspond to a complete sample. It contains only very few low-mass planets, but it has the advantage of showing the existence of a large number of giant planets of long period, which here appear to be separated from the short-period hot Jupiters by a relatively empty zone. From http://exoplanets.eu.

Orbits of all kinds

The Solar System had accustomed us to quasi-coplanar and almost circular orbits around the Sun. This is not the case for many other systems: exoplanets are often located on very elliptical orbits, whose eccentricity may be greater than 0.9 (Figure 5.2). If we plot the eccentricity of the orbits as a function of the mass of the planets, we find that the planets, from the super-Earths to the massive planets, present a great diversity of orbits, from circular to very elliptical. Again, this discovery required a revision of the existing models.

The variety of exoplanets

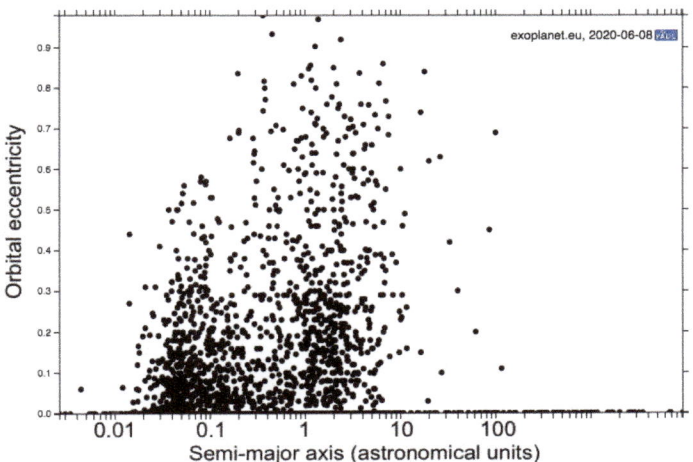

FIGURE 5.2 – Diagram showing the eccentricity of the orbits of exoplanets as a function of their distance from the star. The points of zero eccentricity correspond to a circular orbit, the large eccentricities to very elliptical orbits. The dispersion in eccentricity is very large for all the distances to the star; the same applies to the mass distribution (not shown). In the Solar System, the eccentricities of the planets are all less than 0.1 except for Mercury (Appendix 1). From http://exoplanets.eu.

Many multiple systems

At the time of writing, on a total of about 4270 exoplanets, more than 700 systems comprising at least two planets have been found. The Solar System is therefore not unique! Most of these systems have been detected by transit: their planets therefore evolve in the same plane as those of the Solar System. However, some planets of multiple systems observed by velocimetry have inclined or eccentric orbits.

Examples of multiple systems very different from each other are presented in Appendix 2. One of the most amazing is the system of five planets orbiting the star 55 Cancri; it is particular in several ways. This star, about forty light-years away from the Sun (12 parsecs), is in fact a double system, composed of a star of type K, smaller than the Sun, and of a red dwarf, 55 Cnc B, still less massive, located at about 1000 a.u. of 55 Cnc A. The planetary system evolves around 55 Cnc A, the brightest component of the system. Its five planets were discovered by velocimetry, and one also observed by transit (Figure 5.3). This complex system has been the subject of multiple studies, sometimes contradictory, and could still house other exoplanets.

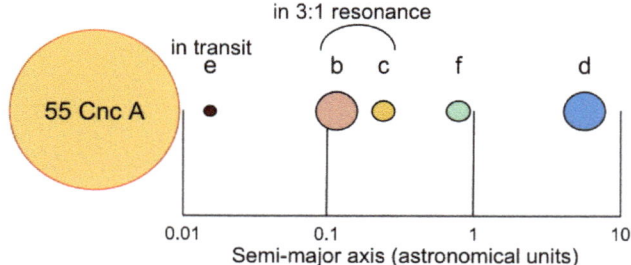

FIGURE 5.3 – The planetary system around the star 55 Cancri A. The planets and the star are not to scale. This system presents an astonishing diversity with two giant planets (b and d), two sub-giants (c and f) and a super-Earth (e) discovered by transit. The orbits are not in the same plane. According to B. Nelson.

Planets around double stars

Are there planets orbiting double or multiple stellar systems? The question deserves to be asked, since we have just mentioned the case (which is not unique) of a system around one of the components of such a system. The question is also interesting because multiple star systems make an important fraction of the stellar population. The answer is yes, exoplanets have been discovered orbiting the center of gravity of some double systems! The first case was discovered in 1993 with the detection of a planet around the binary pulsar PSR-B1620-26. A little more than twenty years later, a planet was detected around a system consisting of a solar-type star and a brown dwarf. Other discoveries of the same type followed. The different types of orbits possible for exoplanets in orbit around a binary system are presented in Appendix 4.

The different classes of exoplanets

We have seen that the radius of the planets detected by transit can generally be determined. Figure 5.4 shows how the exoplanets of different sizes currently detected are distributed according to their period of revolution around the star. Three main classes of exoplanets are clearly distinguished: the cold and warm Jupiters, which occupy the upper part of the diagram, the warm ones being concentrated near the host star; the super-Earths whose radius is between 2 and 4 Earth radii and the mass is of the order of a dozen Earth masses; the Earths, which we will define as having a radius of less than two terrestrial radii, which appear even more concentrated than the super-Earths near the star. Between the Jupiters and the super-Earths, one finds planets of dimensions comparable to those of Saturn or Neptune, in relatively small number.

The variety of exoplanets 63

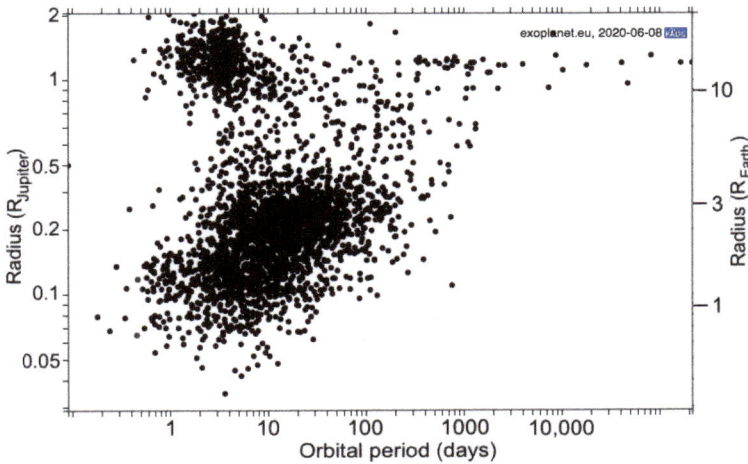

FIGURE 5.4 – Distribution according to their period of revolution of the radius of exoplanets. We can distinguish at the top the hot Jupiters, of radius 1 to 2 $R_{Jupiter}$, close to their star and therefore of short period, and with longer periods, colder planets of dimensions close to that of Jupiter and a few cooling planets very far from their star. Below, super-Earths with a radius between 0.18 and 0.35 $R_{Jupiter}$ (2 to 4 terrestrial radii) and probably rocky planets more similar to the Earth with a radius smaller than 0.18 $R_{Jupiter}$ (2 R_{Earth}), whose period is shorter and which are therefore closer to their star. Between the super-Earths and the Jupiters, a small number of planets of dimensions comparable to Saturn (0.84 $R_{Jupiter}$) or Neptune (0.35 $R_{Jupiter}$). This diagram is heavily biased against long-period planets. According to http://exoplanets.eu.

For planets that have been observed by velocimetry, the mass m is known, or at least the quantity $m \sin i$ and therefore a lower limit of mass if they have not been observed also by transit. There is another method for obtaining the mass in the case of multiple systems: the mutual gravitational interaction between the planets results in irregularities in their motion, and thus in the times of passage in front of the star, and celestial mechanics allows to obtain their mass. However, this method, known as TTV for *Transit Timing Variation*, gives good values of the masses in rare cases only.

We have seen above (Figure 5.1) the relation between the mass obtained by one or the other of these methods and the period of revolution of the exoplanets. It must be borne in mind, however, that neither radii nor masses are generally known accurately, for they depend not only on observations but also on the estimates of the radius and mass of the star which are generally not accurate. We will now study the properties of the different classes of planets that we have identified in Figures 5.1 and 5.4.

Hot and Cold Jupiters

These are the most massive exoplanets, the mass of which can exceed a dozen times that of Jupiter (as we shall see, the boundary between massive planets and more massive brown dwarf stars is uncertain; the International Astronomical Union fixes this limit to 13 masses of Jupiter). When they are close to their star, they are hot and have an evaporating atmosphere, which is observed by spectroscopy as they pass in front the star. It is not surprising that these hot Jupiters were the first exoplanets to be discovered by velocimetry. Many have been also discovered by transit. On the other hand, the probability of transit of planets far from their star, and in particular the cold Jupiters, is low so that most have been discovered by velocimetry. Figure 5.5 shows the distribution of the Jupiters according to their period of revolution, from a complete sample of 63 giant planets discovered by transit and confirmed by spectroscopy with the specialized SOPHIE spectrograph attached to the 1.93 m diameter telescope of the Haute Provence Observatory. Each has been assigned a weight a/M_* in order to correct the result for the transit probability in order to obtain the real distribution. Knowing the number of stars observed by the Kepler satellite, from which this distribution was obtained, we find that at least 10% of stars have giant planets, and that the hot Jupiters are relatively rare.

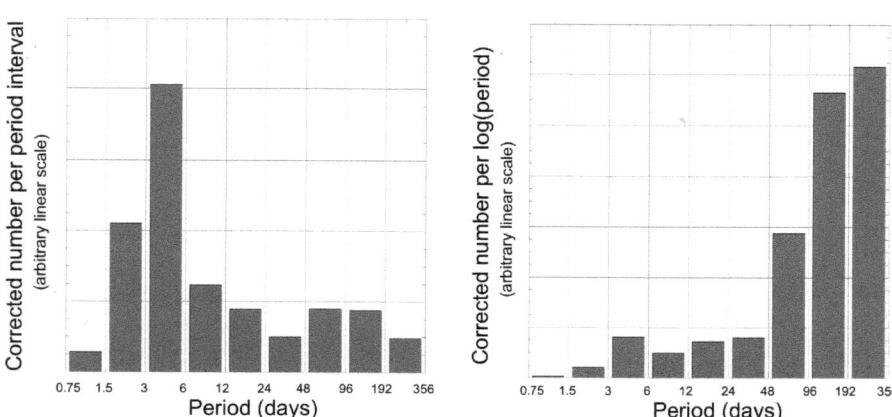

FIGURE 5.5 – Histograms of the distribution of giant planets according to their period of revolution (in logarithmic scale), obtained from a complete sample of planets discovered by transit and confirmed by velocimetry, and corrected for the probability of transit. On the left, the corrected number per linear period interval, and on the right by logarithmic period interval. We show these two diagrams in order to warn of the danger of hasty conclusions that one could deduce from only one of them. Schema by the authors.

Since the mass and the radius of the planets of the complete sample are known, their mean density can be calculated and compared to that of Jupiter (1.33 g/cm³). The result is shown in Figure 5.6, where also almost all the planets whose mass and density are currently known with sufficient accuracy are plotted. The observed correlation between density and mass of giant planets is remarkable. It corresponds to a nearly constant radius (to within 30%) whatever the mass, which is theoretically predicted for planets consisting essentially of hydrogen and helium, as are Jupiter and Saturn. This plot shows that all hot or cold Jupiters are in this case.

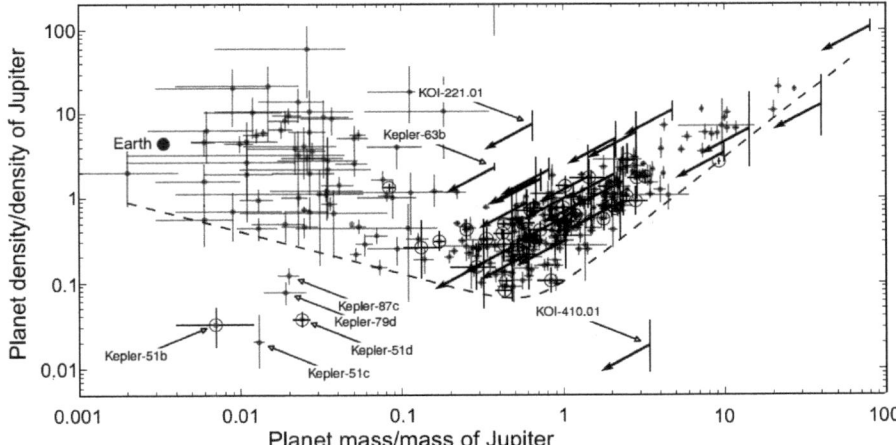

FIGURE 5.6 – The mean density of the planets as a function of their mass. The density is relative to that of Jupiter (1.33 g/cm³), and the mass is in units of mass of Jupiter (logarithmic scales). The arrows indicate planets for which only an upper limit of the mass could be obtained. The circles are relative to the complete sample of giant planets and the points to the other exoplanets, with error bars. The dotted line is an empirical lower envelope of the density for planets of different masses. The planets of very low density are perhaps of a particular nature, but their mass has been determined by the TTV method whose results are generally uncertain. From Santerne et al. (2016).

On average, the giant planets have a larger radius if they are closer to their star, as shown in Figure 5.7. Several mechanisms, which can act simultaneously, can explain the dilatation of these highly irradiated planets. However, some giant exoplanets have a radius of up to 2 $R_{Jupiter}$, which is not yet well understood.

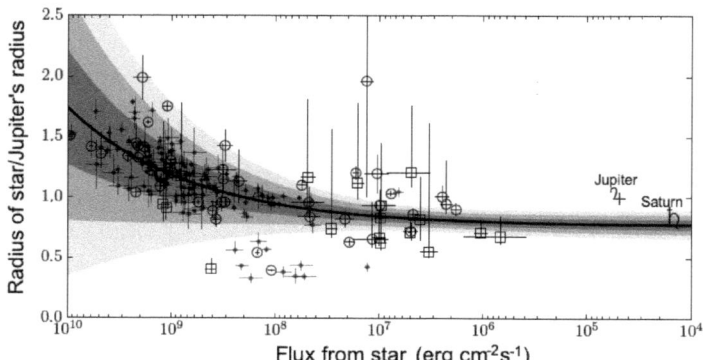

FIGURE 5.7 – Radius of the giant planets according to the energy received from their star. The bold line corresponds to the best fit of the data, and the neighboring areas correspond to the confidence intervals of 1, 2 and 3σ. The position of Jupiter and Saturn is indicated by their symbol. From Santerne et al. (2016).

There is a good correlation between the probability for a star to be surrounded by giant planets and its abundance of heavy chemical elements such as iron (the *metallicity* in astronomical jargon), as shown in Figure 5.8. This correlation was discovered as early as 2001 and is confirmed by recent observations. It means that the giant planets form much more easily in an environment rich in heavy elements. This is explained by the fact that the giant planets originate from a core of heavy elements, of about ten Earth masses, on which the gas of the protoplanetary disk is accreted until it disappears from the disk after a few million years (see Chapter 7). The greater the metallicity in the primordial material, the sooner the nucleus is formed and the more gas it can accumulate, which eases the formation of giant planets.

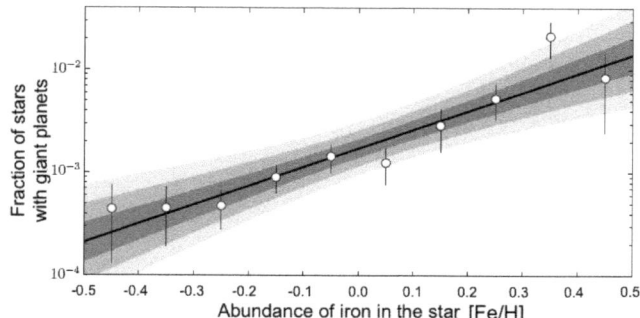

FIGURE 5.8 – Fraction of dwarf stars hosting a giant planet discovered by transit, as a function of their abundance of heavy elements (here iron). Abundance is given as [Fe/H] = log (Fe/H)$_{\text{star}}$ − log (Fe/H)$_{\text{Sun}}$. The bold line corresponds to the best fit of the data and the surrounding areas correspond to the confidence intervals of 1, 2 and 3σ. From Santerne et al. (2016).

Super-Earths and Neptunes

There is not yet a complete (unbiased) sample of planets of mass smaller than that of the Jupiters, observed by transit and verified by velocimetry. However, complete samples have been obtained by velocimetry alone, with the disadvantage that this method gives only $m \sin i$, therefore only a lower limit of the mass. Figure 5.9 shows such a sample, for which all dwarf stars with a mass between 0.6 and 1.6 M_{Sun} (spectral type between M0 and F0) inside a well-defined volume centered on the Sun have been observed. This excludes giant or sub-giant stars, hot stars that are not very numerous, and brown dwarfs. The stars, numbering 822, were observed with the specialized spectrographs CORALIE and HARPS attached respectively to the 1.2 m diameter Swiss telescope and the 3.6 m telescope of the European Southern Observatory, both in Chile. CORALIE gives information on giant planets with period up to 12 years, thus completing the sample discussed in the previous section, and HARPS informs on planets with a lower mass and period less than a hundred days. Some stars have been observed by other means. A total of 155 planets belonging to 102 planetary systems were thus discovered.

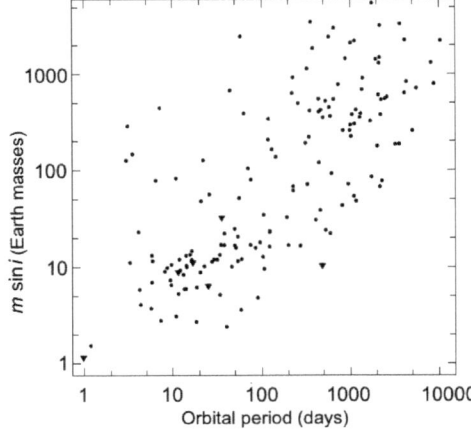

FIGURE 5.9 – A complete sample of planets observed by velocimetry. Compare with Figure 5.1. One notices the large number of giant planets of long period compared to the hot Jupiters of short period. There is a significant bias against low-mass planets. From Mayor, M. et al. (2011).

This sample provides valuable insights on planets of lower mass than Jupiter. It can be seen (Figure 5.10) that their number increases considerably for masses of less than about 30 M_{Earth}. They are super-Earths and planets of mass comparable to that of Neptune or Uranus. Most of them have periods smaller than 100 days (Figure 5.11), which is not the case for Jupiters (see Figure 5.5 right). Figure 5.1 suggests that they are on average closer to the star as their mass is smaller. These planets are very common: they exist around 50% of the stars, which adds to the 10% of stars that possess giant planets.

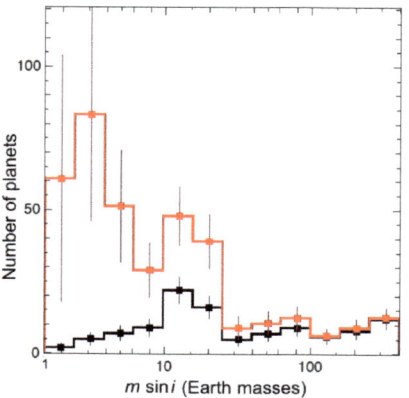

FIGURE 5.10 – Distribution of the masses of the planets of the sample of Figure 5.9, by logarithmic interval of mass. In black, the number observed. In red, the number corrected for detection biases, with error bars. Note the concentration of these objects at masses below 30 M_{Earth}: they are Super-Earths or Neptunes (17 M_{Earth}), with perhaps the separation at 10 M_{Earth} suggested by Figure 5.4. The more massive planets of this diagram, in smaller numbers, are comparable to Saturn (95 M_{Earth}), then we reach the Jupiters (318 M_{Earth}). From Mayor, M. et al. (2011).

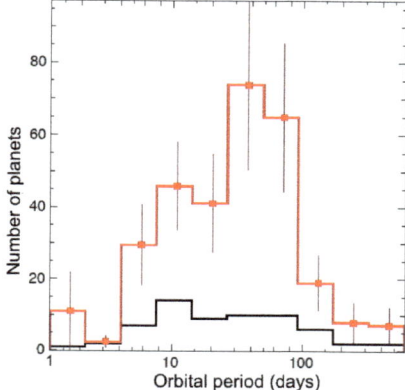

FIGURE 5.11 – Distribution of the planets in the sample of Figure 5.9 and of mass such that $m \sin i < 30$ M_{Earth}, by logarithmic interval of period of revolution. These are super-Earths and Neptunes. In black, the number observed. In red, the number corrected for detection biases, with error bars. The vast majority of these planets have periods of less than 100 days. From Mayor, M. et al. (2011).

There is, however, a lack of information for planets of mass close to that of the Earth, which are perhaps even more numerous. One can therefore think that there are at least as many planets as stars in the Galaxy, about one hundred billion, and maybe even more.

Figure 5.6 shows that the mean density of these planets increases as their mass decreases. With its 5.52 g/cm^3, the Earth is in the middle of the least massive planets of this diagram, which are entirely rocky. Neptune and Uranus, which are partly composed of ice, have respective densities of 1.64 and 1.32 g/cm^3, also in the average of the planets of their mass, which may therefore have a similar chemical composition.

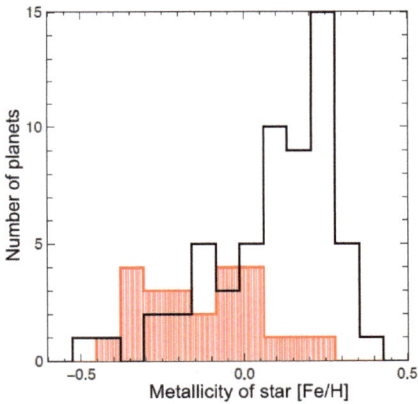

FIGURE 5.12 – Distribution of the metallicity of host stars of super-Earths and Neptunes ($m \sin i$ <30 M$_{\text{Earth}}$), in red, compared to that of host stars of giant planets, black contour. The metallicity is [Fe/H] = log (Fe/H)$_{\text{star}}$ − log (Fe/H)$_{\text{Sun}}$. From Mayor, M. et al. (2011).

The abundance of heavy elements (metallicity) of the host star, which is that of the primordial cloud from which the star and the planets formed, does not seem to play a large role in the formation of super-Earths and Neptunes, unlike what happens for the giant planets that depend much on it. This is illustrated in Figure 5.12, which shows that the host stars of giant planets and those surrounded by planets with $m \sin i$ < 30 M$_{\text{Earth}}$ show very different metallicity distributions. These latter planets do not accumulate the surrounding gas and have much more time to grow than the giant planets making them less sensitive to metallicity.

Earths and Habitable Planets

A question has obsessed humanity since its origin: are we alone in the Universe? Now that we know that we are surrounded by tens of billions of exoplanets, in our Galaxy only, the question is more than ever topical. How can we approach it? We do not know how life has emerged on Earth, but we have been able to identify some factors favorable to its emergence (see Chapter 10): the presence of an energy source and the presence of liquid water on a solid surface, in order to allow the development of an increasingly complex chemistry within a liquid medium. In the absence of other information, astronomers in search of extraterrestrial life favor the environments in which the conditions of appearance of life on Earth are supposed to be met. The first condition will therefore imply a rocky planet around a star that will be the source of energy. This imposes a constraint on the mass of the object. Indeed, models of planetary formation (see Chapter 7) teach us that beyond a mass of about ten Earth masses, the gravity field of the object is sufficient to capture the gas of the surrounding protostellar nebula: this is how the giant planets form. The habitable planets must therefore have a mass below this threshold of ten Earth masses: these are the super-Earths, and the Earths themselves. Note in passing that in the quest for extraterrestrial life, identifying a planet with the size of the Earth is not a necessity: it is enough that it is a rocky planet, which can be a super-Earth. Presently, there are more than 200 known exoplanets with a mass below the threshold of ten Earth masses... the choice is therefore wide.

Another criterion must be met: the exoplanet must be able to have liquid water on its surface. Its equilibrium temperature must therefore lie between 0 °C and 100 °C, i.e. 273 K and 373 K (the upper limit is not an absolute constraint because it depends on the surface pressure of the exoplanet). The researchers studied closely the conditions under which liquid water could remain for a sufficient lifetime. Following the pioneering work of the American James Kasting, the notion of "habitability zone" emerged (Figure 5.13): this is the region where the equilibrium temperature allows the water to be in liquid form. For stars similar to the Sun, it is centered around 1 a.u. It is closer to the star if it is less luminous (which is the case of the K and M stars which represent the vast majority of the stellar population) and farther away for the massive and luminous stars O, B and A. We will see later (Chapter 10) how current research attempts to identify the exoplanets that are likely to be in this famous zone of habitability.

The variety of exoplanets

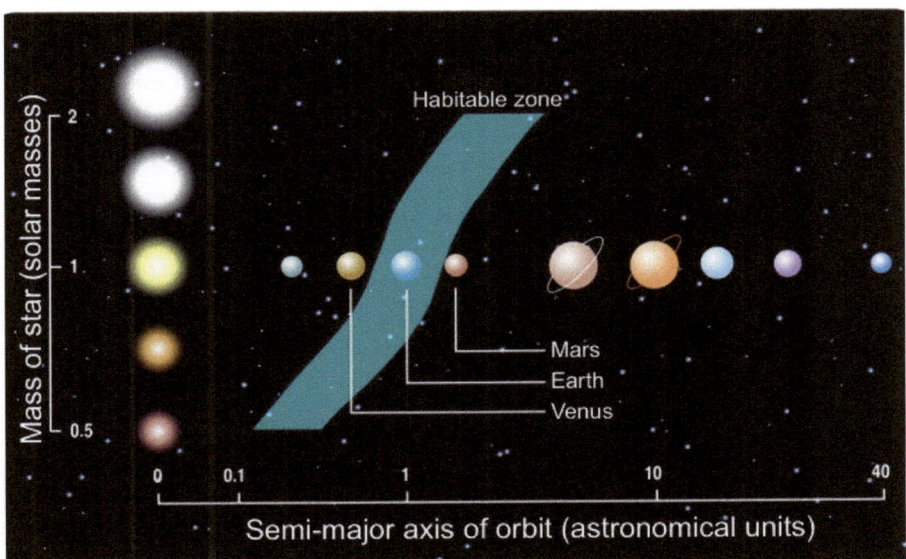

FIGURE 5.13 – Distance and width of a star's habitable zone according to the mass of the star. The diagram shows the position of the planets of the Solar System to illustrate the case of a star of a solar mass. The width of the habitable zone may vary according to the authors, according to the criteria chosen to define the limits (see Chapter 10). From http://www.astronoo.com/en/articles/exoplanete-kepler-22b.html.

It is important to note that the "habitability zone", defined in relation to the presence of liquid water on the surface of the object, is only a criterion for bringing together certain conditions favorable to the emergence of life. It does not imply that life will appear in the habitable zone: we have no idea as to the probability of such an event. On the other hand, we cannot exclude the possible presence of life in very different environments, for example in the oceans of liquid water that are probably present inside certain external satellites of the Solar System: Europe, Enceladus, may be also Ganymede. It should also be noted that our current knowledge of exoplanets gives us only a very approximate idea of their equilibrium temperature (see box). In order to conclude that liquid water is present on an exoplanet, we would have to know its atmospheric composition, which is largely unknown for now, except for a few giant planets (see Chapter 8).

Box: a simple model to characterize exoplanets

Let us first estimate the equilibrium temperature T_p of an exoplanet. Let R_* and T_* be the radius and the effective temperature of the star (approximate values of these quantities can be found in Appendix 3 as a function of the spectral type of the star). The flux received by the planet at the distance a per unit area perpendicular to the direction of the star, is $\sigma T_*^4 R_*^2/a^2$, where σ is the Stefan-Boltzmann constant. The planet only absorbs the fraction $1-A$ of this flux, A being its albedo, on a surface of πR_p^2, R_p being its radius. At thermal equilibrium, if the planet rotates rapidly, it emits by its entire surface an energy σT_p^4 per unit area, that is to say a total of $4\pi R_p^2 \sigma T_p^4$. So we have:

$(1-A)\pi R_p^2 \sigma T_*^4 R_*^2/a^2 = 4\pi R_p^2 \sigma T_p^4$, or, simplifying:

$(1-A)T_*^4 R_*^2/a^2 = 4 T_p^4$, hence

$T_p = (1-A)^{1/4} (R_*/2a)^{1/2} T_*$, and numerically, if R_* is in solar units and a in astronomical units:

$T_p = 0.0482 (1-A)^{1/4} (R_*/a)^{1/2} T_*$.

If the planet is in rotation synchronous with its revolution, which is generally the case if its period of revolution is less than a few tens of days (the value of this limit depends on the nature of the planet), it emits in principle by half its surface only, and the temperature of this illuminated half is:

$T_p = (1-A)^{1/4} (R_*/2^{1/2}a)^{1/2} T_*$, or numerically $T_p = 0.0573 (1-A)^{1/4} (R_*/a)^{1/2} T_*$.

However, there is a partial redistribution of temperature by the winds, and the mean average temperature is intermediate between that of the two cases.

T_p is not very sensitive to the exact value of the albedo A. But A may be very different from one planet to another and is difficult to evaluate. For example, the bolometric albedos (i.e. the fraction of the reflected solar energy) of Mercury, Venus, Earth, Mars, and Jupiter or Saturn are respectively 0.12, 0.75, 0.31, 0.25 and 0.34. If our planets were illuminated by the predominantly infrared radiation of a relatively cold M star, their albedos would be different: for example, a rocky planet illuminated by such a star must have an albedo close to 0. The albedo of exoplanets can in principle be determined near the secondary transit, but this would require that their flux be measured not only in the infrared but also in the visible, which is currently not possible.

Moreover, the temperature can be increased by the greenhouse effect: for example, the greenhouse effect due to H_2O and incidentally to CO_2 and CH_4 brings the temperature of the Earth from 254 K to 288 K, while the CO_2 raises the temperature of the surface of Venus from 328 K to 730 K.

Let us now compare the estimated temperature of the exoplanet to the condensation temperature of water vapor at low pressure, i.e. 230 K. If it is lower, the planet is an icy object; if its mass is greater than 10 Earth masses, it has been able to accrete hydrogen and helium and is therefore a giant planet. If the temperature is above about 230 K, we are dealing with a rocky planet, possibly habitable if it is not too massive and has an atmosphere and water, or a hot Neptune or Jupiter if it is massive.

The variety of exoplanets

If the temperature is high and the pressure low, the chemical equilibrium predicts that the atmosphere is composed mainly of CO, N_2, H_2O and CO_2 formed by CO + $H_2O \rightarrow CO_2$ + H_2 (see Chapter 8). If the temperature is low and the pressure high, CH_4 and NH_3 and optionally H_2O are found, in addition to H_2 and He. The following diagram illustrates the classification of exoplanets according to their temperature and mass.

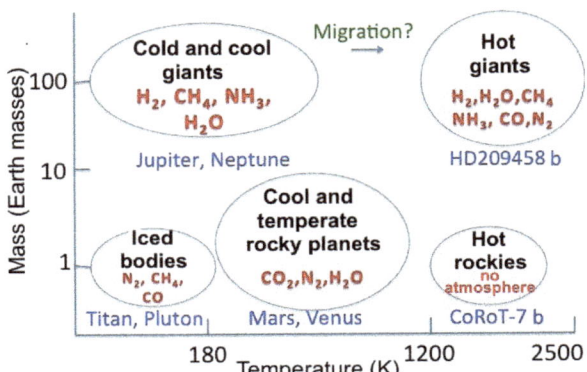

FIGURE 5.14 – Classification of exoplanets according to mass and temperature. The young planets in the process of cooling are missing in this diagram.

Bibliography

Kasting, J.F., Whitmire D.P. & Reynolds R.T. (1993) Habitable Zones around Main Sequence Stars, *Icarus* 101, 108.

Mayor, M. et al. (2011) The HARPS search for southern extra-solar planets XXXIV. Occurrence, mass distribution and orbital properties of super-Earths and Neptune-mass planets, arXiv:1109.2497

Santerne, A. et al. (2016) SOPHIE velocimetry of Kepler transit candidates. XVII. The physical properties of giant exoplanets within 400 days of period, *Astronomy & Astrophysics* 587, A64

Chapter 6
The birth of stars and protoplanetary disks

Protostars, jets and disks

We have briefly described in Chapter 1 the model of formation of planetary systems presented initially by Kant and Laplace, which seems to agree with what we observe of the Solar System. Let us recall its main lines (Figure 6.1).

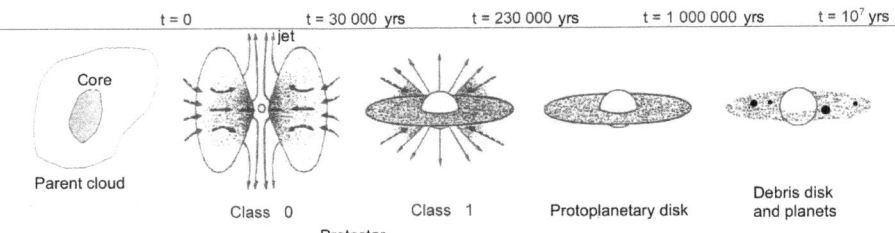

FIGURE 6.1 – The stages of formation of a star and its circumstellar disk. Under the effect of gravity, a part of an interstellar cloud collapses on itself, faster in its central regions, forming a dense core. When the heat generated by the collapse can no longer be evacuated from this core, the temperature rises until nuclear reactions occur, then the star lights up (t = 0). While matter continues to be accreted by the star, a bipolar jet evacuates some of the material with its angular momentum, which slows down the rotation of the star (Class 0 proto-star). Then, a circumstellar disk forms, while the jet fades (Class 1). The jet stops and the accretion decreases; planets begin to form in the disk that can now be called a protoplanetary disk, the dust and gas of which are finally expelled by the wind and the radiation of the star. After a few million years, there remains only a young star and some planets in a disk of debris. Diagram of the authors, after Philippe André.

Stars and their planetary systems are formed by the gravitational collapse of an interstellar cloud of gas and dust, such as that associated with the Orion Nebula, known as a "star nursery" (Figure 6.2). Observations indicate that the coldest clouds, where the gas is essentially in the form of molecules, the most abundant of which is by far the hydrogen molecule H_2, are the sites of the formation of stars of mass comparable or lower than that of the Sun. The giant clouds, like that of Orion, are generally warmer and produce stars of all masses. However, the theory of star formation is difficult because of the complexity of the structure and dynamics of molecular clouds and also because the physical phenomena involved are multiple and often still poorly understood, like turbulence that most likely exists in these clouds. The role of the magnetic field, although certainly important, is unclear and controversial. Nevertheless, it appears that the cloud must fragment during its collapse, which corresponds to the observation since the stars are generally formed in groups.

FIGURE 6.2 – The Orion Nebula and its cluster of young stars, seen in the infrared. This composite false color image was obtained at 1.24 µm (blue), 1.65 µm (green) and 2.16 µm (red) wavelengths, with one of the 8-m telescopes of the VLT at the European Southern Observatory (ESO). The nebula produces the diffuse radiation: its gas is ionized by the ultraviolet emission of the four bright and hot stars in the center, which form the Orion Trapezium. At the back of the nebula is an optically invisible molecular cloud. Beneath the surface of this cloud is a rich cluster comprising about 1000 stars formed in three steps during 1 million years: these stars are only revealed by their infrared emission. © ESO.

The collapse compresses the gas from these fragments, which heats them, like the gas compressed by a bicycle pump. The resulting heat is evacuated by the emission of molecular lines, and also by the radiation of dust in the far infrared. But there is a time when the thickness of the material is such that it becomes opaque, even in the far infrared and in millimeter radio waves, so that the heat can no longer be evacuated: the medium then continually heats up while its density becomes very large. Meanwhile, the dense core of the cloud continues to grow due to the fall of the surrounding matter. If the mass of the nucleus is sufficient, more than about 0.07 times that of the Sun, thermonuclear reactions of fusion of hydrogen into helium start and the star is born. If it is smaller, we are dealing with an aborted star, a brown dwarf.

The theory predicts that there is a lower limit to the mass of the cloud fragments that are likely to collapse, a limit that is about one hundredth of the mass of the Sun, ten times the mass of Jupiter. Thus, brown dwarfs can form directly, like ordinary stars, from fragments of molecular clouds, but not the planets which are necessarily produced in another way, except perhaps the most massive.

If the protostellar fragment rotates rapidly on itself, it can generate a double or multiple star. If its rotation is moderate, its collapse can only occur along the axis of rotation, so that the object turns into a rotating disk with a central condensation. However, in our Solar system and in the exoplanetary systems, we observe that most of the angular moment is not in the central star but in the surrounding planets. Therefore, there must have been an outward transport of the angular momentum of the central condensation. This has long been a problem because the mechanisms at work in the protoplanetary disks (turbulence, magnetic field) seem insufficient. The solution, unexpectedly, involves jets of material along the axis of rotation of the star, which appear to emerge from all the stars in formation (Figure 6.3). We are beginning to understand how these jets are formed: the loss of angular momentum occurs through a subtle mechanism which uses the magnetic field of the central parts of the circumstellar disc to pump its rotational energy and propel a bipolar jet along the rotation axis.

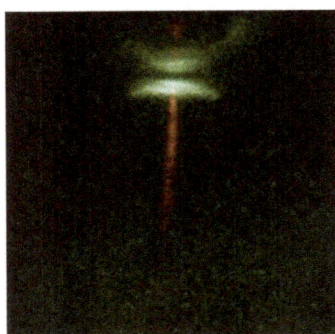

FIGURE 6.3 – The disk seen edge-on and the jet of the star in formation HH 30, observed with the Hubble space telescope (true colors). The dust of the disk diffuses the light of the star, while the gas which forms the jet emits mainly in the Hα line of hydrogen, hence its red color. The star itself is invisible because the presence of dust makes opaque the middle zone of the disk, hence the central dark band. © Hubblesite.

What happens in the disk? A large part of it falls on the central star after its angular momentum has sufficiently diminished (the jets disappear at this stage), and only a small part of its initial mass remains. It is there that the planetesimals form, which are later mostly gathered in elements of kilometric size, the planetoids, and then in planets (see Chapter 1). In the internal regions, the gas has disappeared and the planetoids are made up of rocky elements (silicates, metals, metal oxides, etc.). Beyond a distance from the star sufficient for water and possibly other molecules to remain in ice form (the circle having this radius is called the *ice line*), the planetesimals incorporate them. We will clarify these notions in the next chapter.

Let us stop here and see what the disks are like when planets are formed.

The protoplanetary disks

The last two decades have seen numerous observations of disks around young stars, disks that were discovered in 1983 by their infrared emission, as we saw in Chapter 1. This discovery was not unexpected: the presence of disks around the forming stars had been anticipated long before (this is the basis of the ideas of Kant and Laplace), and as early as 1974 the theory of circumstellar disks was produced. Given the large distance from us of the stars around which they gravitate, these disks have very small angular diameters, so that an excellent spatial resolution is necessary in order to obtain images, which implies the use of very powerful observation means: in the visible and near infrared, the Hubble Space Telescope, in the near or middle infrared, a telescope equipped with adaptive optics and in the radio domain, millimeter and submillimeter interferometers. Figure 6.4 schematically shows the different emissions of protoplanetary disks, from which we can deduce what the different types of observations can bring.

In a protoplanetary system, the bulk of the mass is in the star, and less than 1% in the disk. This excludes that the planets are formed by gravitational collapse of parts of the disk, an idea that had been advanced by Cameron, but which requires that the disk has a mass comparable to that of the star. The planets, clearly, are not born of the same physical mechanisms as the stars. Moreover, we observe that the mass of the disk is more or less proportional to that of the star. Finally, it is estimated that 99% of the disk is in the form of gas, and therefore only 1% in the form of solids.

The dimensions of the disks are very variable, their radius ranging from 5 to 200 a.u. Their thickness is difficult to measure, but it seems that some are rather thin and others are thicker. Even if some uniform disks seem to exist, the density of matter generally increases when we approach the star. Indeed, over time, the gas and the dust which it entails tend to concentrate towards the interior of the disk. On the other hand, the most central parts are not very dense because of the continuous accretion of the material of the

The birth of stars and protoplanetary disks

disk on the star and the dispersion of the gas due to the ultraviolet radiation of the star and the wind it emits. A part is ejected perpendicular to the disk and falls back on the star. But most of the dust, which has very rapidly agglomerated into planetesimals, then planetoids and planets, escapes these phenomena. Finally, after a few million years, the gas and dust disappear completely from the disk and only solid debris remains.

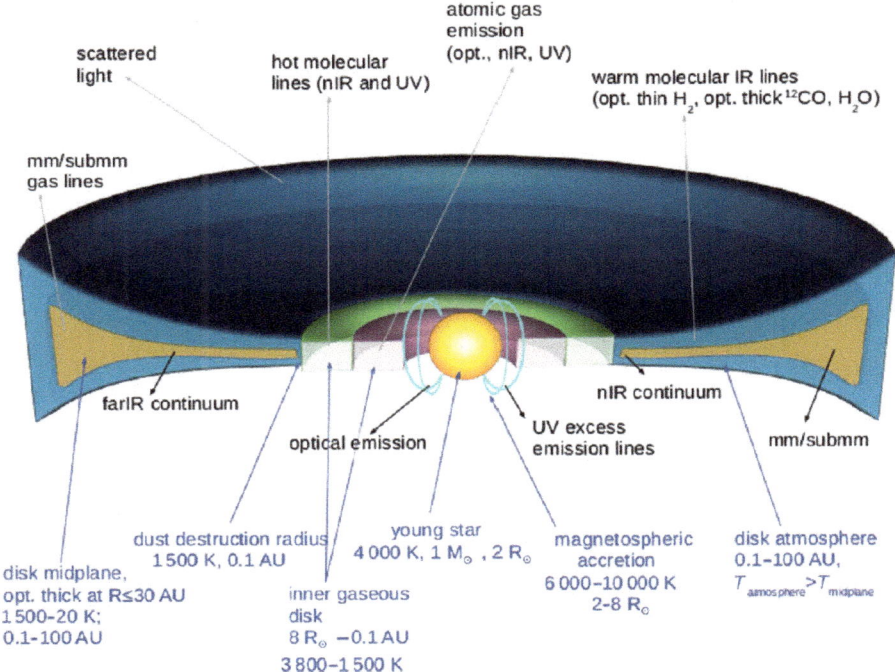

FIGURE 6.4 – A schematic section of a protoplanetary disk with its different emissions. The dimensions are only indicative, as they vary from one disk to another. From Sicilia-Aguilar, A. et al. (2016).

The formation of planets can profoundly alter the structure of the disk. Very detailed images of protoplanetary disks were recently obtained at millimeter and submillimeter wavelengths by the ALMA interferometer in Chile. Figure 6.5 shows the disk surrounding the star HL Tauri, a proto-star somewhat more massive than the Sun and only 1 to 2 million years old. Its distance is 460 light-years.

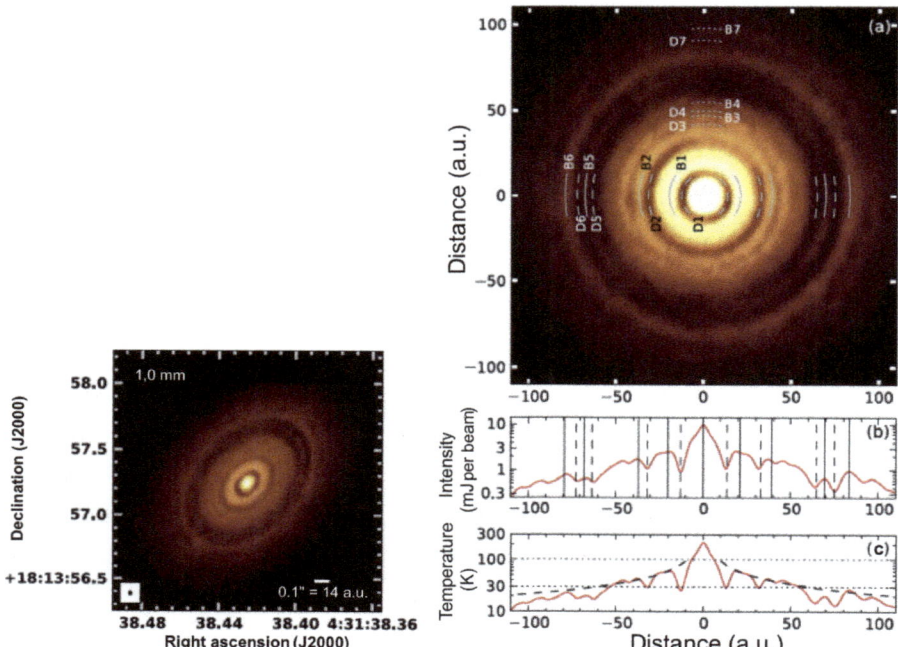

FIGURE 6.5 – The protoplanetary disc surrounding the star HL Tauri. On the left, the emission of dust as observed with the ALMA interferometer at 1 mm wavelength. The angular resolution (black ellipse in the small white square) is 30 × 19 milliseconds of degree. The dark rings correspond to increases in dust density, hence in optical thickness. On the right, the same image, rectified, and two sections giving the intensity (b), and the temperature of the dust (c, the curve in broken lines gives the real temperature, since the emission is only that of the black body when the optical thickness is large). From Brogan, L. et al. (2015) *Astrophysical Journal Letters* 808, L3.

In this remarkable image, it is observed that the emission of the disk is not uniform and presents an alternation of bright and dark rings. Here, what is observed is the thermal radiation of the dust, heated by the young central star. The disk also contains gas revealed by the emission of molecular lines, but it is not possible to obtain its detailed structure because of contamination by the gas ejected by the star.

The image of the disk surrounding TW Hydrae, another young star, but somewhat older than HL Tauri (10 million years), is particularly detailed because this star, at 180 light-years, is closer than HL Tauri. Details as fine as 1.6 a.u. are seen (Figure 6.6).

These rings may betray the presence of planets in some cases. Indeed, the mass of the planets, if any, is not negligible compared to that of the disk, and they interact gravitationally with it, generating in particular annular or spiral structures. The disk is then called a *transition disk*. Three examples are shown in Figure 6.7.

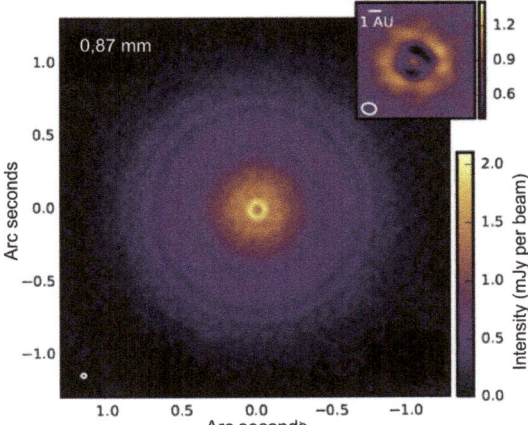

FIGURE 6.6 – The emission of dust at 0.87 mm wavelength observed with ALMA around the young star TW Hydrae, rectified view. The coordinates are relative to the center of the disk. The small white circle indicates the angular resolution, 30 milliseconds of arc. To the top right, enlarged view of the central part with a slightly better angular resolution, 24 × 18 milliseconds (white ring). Note the alternation of dark and brighter rings, the central hole and the emission of dust just around the central star, which is not visible here. The dark rings are less markedly pronounced than for HL Tauri (Figure 6.4). © ESO, according to Andrews, S.M. et al. (2017).

FIGURE 6.7 – Annular and spiral structures in transition disks. From left to right, the disks around the stars HD 97048, SAO 206462 (also called HD 135344B) and RX J1615, observed with the SPHERE instrument of the ESO VLT at 1.6 μm. These stars, not visible in these images, are respectively about 2.5, 8 and 2 million years old. The structure is most likely due to the gravitational action of planets. However, these planets have not yet been observed directly in any of the cases. (© ESO, de Boer, J. et al., Ginski, C. et al., and Stolker, T. et al.).

The ice lines in protoplanetary disks

The images we have just shown are those of the dust in the disks, revealed by its thermal radiation in millimeter or submillimetric waves (Figures 6.5 and 6.6) or by the diffused stellar light in the near infrared (Figure 6.7). But they do not give the distribution of the gas. Near the star, the temperature is high enough so that all the non-refractory molecules of the disk are in the gaseous state. However, when at larger distances from the star, the temperature of the dust becomes sufficiently low so that the least volatile molecules can condense there: first water vapor when the dust is between 100 and 130 K (the exact temperature depends on the pressure), then ammonia, methane and carbon monoxide at lower and lower temperatures. The transition zone between the gas phase and the condensed phase is called the *ice line*.

The first observed one was the CO ice line, where the dust temperature is about 20 K: it is very far from the star, which made it possible to observe it with the angular resolution of ALMA. This observation was made on the disk surrounding the young star HD 163296 (Figure 6.8), both directly by observing the distribution of the CO molecule, or rather its $C^{18}O$ isotopomer since the lines of the main molecule $C^{16}O$ are very optically thick, and indirectly by observing other molecules. Indeed, the molecular ion N_2H^+ is destroyed in the presence of gaseous CO by the reaction $N_2H^+ + CO \rightarrow N_2 + HCO^+$, and its observation indicates the absence of CO. Figure 6.8 shows that N_2H^+ and gaseous CO do not coexist. The HCO^+ molecule and its isotopomer DCO^+ coexist with CO, which is not surprising since they form from CO. The observations in Figure 6.8 show that CO is gaseous up to a distance from the star of 90 a.u., and solid beyond. It is therefore the radius of the CO ice line.

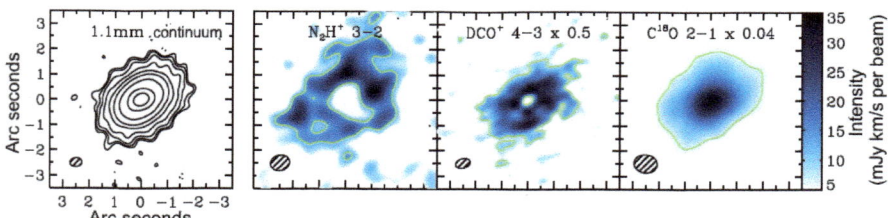

FIGURE 6.8 – Images of the disc surrounding the very young A1V star HD 163296, distant of 119 pc, obtained with ALMA. The ellipses indicate the angular resolution. From left to right, the 1.1 mm wavelength continuum traces dust; the molecule N_2H^+ molecule disappears in the central regions where CO is found in gaseous form; the DCO^+ molecule, isotopomer of HCO^+, has a distribution close to that of CO but is deficient in the central regions because the deuterated molecules are scarce at relatively high temperatures; the CO distribution is approximately given by that of its less optically thick isotopomer $C^{18}O$, to the right. From Qi et al. (2015) *Astrophysical Journal* 813, 128.

The birth of stars and protoplanetary disks

More fundamental is the ice line of water, which is the most abundant molecule after H_2 in the interstellar medium. In fact, the planets which form farther from the star than this line contain a large quantity of ice, whereas those which form closer are made only of refractory elements. Unfortunately, the H_2O ice line is most often rather close to the star, about 3 a.u. in the case of a star of solar luminosity, and water vapor at a temperature of the order of that of the ice line can be detected only in submillimeter waves, and only from space, due to the difficulties caused by the water vapor in the Earth's atmosphere. Although the Herschel satellite has actually detected water vapor in protoplanetary disks, its angular resolution was not sufficient to map them.

However, it has been possible to determine indirectly with ALMA the position of the H_2O ice line in the disk surrounding the very young star V883 Orionis (Figure 6.9), due to a particular circumstance: the star is currently undergoing a very large burst of brightness, reaching 400 times that of the Sun, due to an important fall of circumstellar material on its surface. The dust is heated much farther from the star than usual, so that the ice line is very far away, beyond the orbit of Neptune in the Solar system, and becomes observable with ALMA despite the large distance of the star (1300 light years). As many young stars pass through such a stage, this observation complicates the problem of the formation of giant planets.

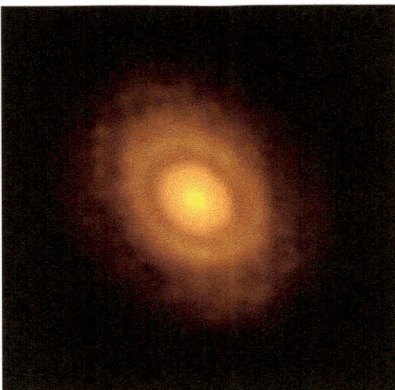

FIGURE 6.9 – The thermal radiation of the dust of the protoplanetary disk around the star V883 Orionis, observed with ALMA near the frequency of 225 GHz. The angular resolution is 0.03 arcsecond (12 a.u. at the distance of the star). The H_2O ice line coincides with the inner dark ring, whose half-axis is about 40 a.u., where the measured dust temperature is about 105 K. © ALMA (ESO / NAOJ / NRAO / L. Cieza).

Planet-disk interactions

Once a planet has formed, it interacts gravitationally with the disk, and this may force it to move inwardly or outwardly: this phenomenon is called migration. Some theoreticians had already developed in 1980 the corresponding mathematical formalism, which corresponded to different astrophysical objects, such as Saturn's rings and their interaction with the planet's satellites, or the disk of our Galaxy. But this phenomenon was neglected for a long time and it has only undergone significant new developments after the discovery of the first exoplanet in 1995 and the finding of giant exoplanets close to their star.

What happens when a planet is born in a disk of gas and dust is illustrated by numerical simulations like that of Figure 6.10. One-arm spiral structures are formed inside and outside the radius at which the planet gravitates, and this is probably the origin of what is seen in a late stage in some transition disks (see Figure 6.7).

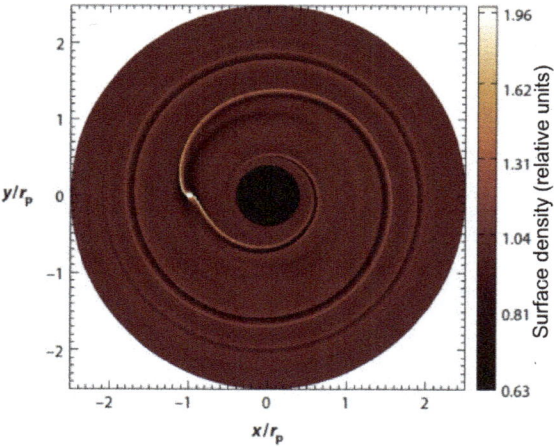

FIGURE 6.10 – Numerical simulation of the gravitational action of a planet on a protoplanetary disk, here supposed for simplicity to have initially a uniform surface density. The figure shows the appearance of the disk after 5 revolutions of the newly formed planet around the central star. The planet and the disk rotate counterclockwise. The gravitational perturbation generates an internal spiral arm; the angular velocity of rotation of the disc increasing inwards, this arm precedes the planet and winds up progressively. There is also the formation of an external spiral arm, this time following the star since angular velocity decreases towards the outside. The color scale on the right indicates the surface density on the disc. From P. Armitage, in Kley, W. & Nelson, R.P. (2012).

The spiral density waves created by the planet in the disk correspond to a transfer of angular moment and thus affect the motion of the planet. The inner spiral arm draws the planet by increasing its angular momentum, as the rotation is faster if one approaches the star; this tends to make it migrate outwards. Conversely, the outer arm brakes it, leading to an inward migration. The net effect is the difference between these two processes, and depends on the details of the geometry and physics of the disk. The computation is complicated, even if the planet has a low mass (say less than ten times the mass of the Earth) and does not significantly disrupt the disk structure (the so-called Type I migration). If the disc has viscosity, which may be the case, numerical simulations must be used. The mechanism is very efficient: the calculation shows that for a very young planet of terrestrial mass gravitating at 1 astronomical unit around a star of solar mass situated in what we imagine to be a pre-solar disk, the characteristic migration time would be of the order of 100,000 years, which is extremely short on the astronomical scale. In order to form planetary systems, various solutions have been proposed to lengthen this time; but this problem is often slipped under the carpet!

If the planet is massive, at least thirty times the mass of the Earth, its gravitational interaction with the disk becomes very strong, thus also the transfer of angular momentum. If shock waves occur or if energy can be dissipated by a high viscosity of the material of the disk, then the angular momentum can simply be transferred to the neighboring matter of the planet without a spiral wave being generated. Then, the matter that gravitates in orbits close to that of the planet but closer to the star loses angular momentum for the benefit of the planet and thus approaches the star. Conversely, the material located farther from the star than the planet is accelerated and moves away from the star. The planet thus opens a circular groove around its trajectory, which we will call a gap. This gap is more or less wide and deep according to the mass of the planet and the viscosity of the disk. The planet is then blocked in this gap and follows the dynamic evolution of the viscous disk, which is generally a gradual accretion by the star: it therefore slowly migrates towards the star. This is called the Type II migration.

If the protoplanetary disk is particularly massive, the exchanges of angular momentum become very large, and one can obtain the very rapid migration of a large planet, which might end up falling on the star. This is called the Type III migration.

Figure 6.11 shows an example of the conditions under which these different types of migrations occur, relative to a particular disk model.

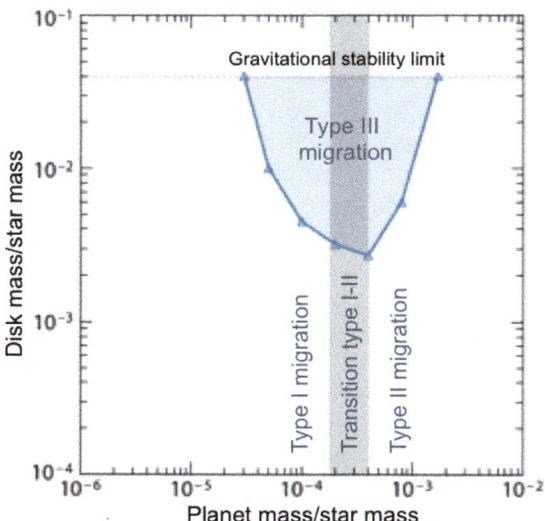

FIGURE 6.11 – Conditions for the different types of planetary migration, as a function of the ratio between the mass of the planet and that of the star, and the ratio between the mass of the disk and that of the star, for a disk with an average viscosity. Disks with mass greater than 0.04 times the mass of the star are not gravitationally stable. It should be noted, however, that these conditions are only valid for a particular disk model, the validity of which is not proven. From Kley, W. & Nelson, R.P. (2012).

To further complicate the situation, we must realize that the mass of the giant planets increases with time, because their nucleus gradually accumulates gas from the disk. Thus the planet can undergo a rapid migration of type I at the beginning, when still not very massive, but when it grows above 30 Earth masses, the migration becomes of type II, less rapid and almost always towards the inside. When its mass reaches that of Saturn, about 100 terrestrial masses, a rapid migration of type III can take place if the disk is massive. But it will not last very long and we will soon return to a slower inward migration of type II, during which the planet has dug a deep gap around its orbit.

We did not mention what happens in a turbulent disk, nor did we describe the evolution of the eccentricity and the inclination of the orbit of the planets on the plane of the disk. In general, these two parameters tend to decrease with time, the orbit becoming more and more circular and in the plane of the disk due to multiple collisions therein. The fact that many orbits of exoplanets are eccentric is almost necessarily explained by the existence of several planets around the star which interact gravitationally with each other, greatly perturbing their respective motions.

If there exist around a star two massive planets whose orbital motion is in resonance, the outer planet making two revolutions, for example, while the inner planet makes three (this is called resonance 3:2), these planets dig into the disk a common gap encompassing their two orbits. The interaction with the disk then takes on a new aspect. If the distant planet is less massive than the planet closer to the star, the transfer of angular momentum with the disk to the outside of the gap is less important than the inward transfer, so that the ensemble of the two planets, which stay in resonance, gain angular momentum: they will therefore move away from the star, whereas for a single planet, the type II migration would have been inwards. This phenomenon is not anecdotal: we will see it at work in the history of the formation of the Solar System.

In the case of HL Tauri (Figure 6.5), various arguments suggest that the main lacunar rings, which are in fact slightly elliptical, are excavated by massive planets. It is striking is that several of these rings resonate with each other: the periods of revolution at the radii of the four inner dark rings D1:D2:D3:D4 are in the 1:4:6:8 ratios, and there are also resonances between the dark and bright rings. Such resonances are observed in 30% of the multiple exoplanetary systems. They appear to be a natural consequence of the dynamics of planetary systems, for example in the planetary systems of TRAPPIST-1 and HR 8799 (Appendix 2). It remains, in the case of HL Tauri, to observe the planets themselves.

The case of TW Hya (Figure 6.6) is quite different: the rings are much less visible than for HL Tauri, and it is not certain that they are due to the action of planets; if so, the mass of these planets must be relatively small. The central hole might correspond to the disappearance of dust following the formation of planets. But there is still much to be done to understand the remarkable structure of this protoplanetary disk... and of the others.

Bibliography

Formation of stars and disks:
André, P., Ward-Thompson, D. & Barsony, M. (2000) From Protostellar Cores to Protostars, the Initial Conditions of Star Formation, in *Protostars and planets, IV*, ed. Mannings, V. Boss, A.P. & Russell, S.S., Tucson, University of Arizona Press, 59, http://cdsads.u-strasbg.fr/abs/2000prpl.conf...59A.
Lequeux, J, (2013) *Birth, evolution and death of stars*, EDP Sciences & World Scientific.

Properties of protoplanetary disks:
Armitage, P.J. (2011) Dynamics of Protoplanetary Disks, *Annual Review of Astronomy & Astrophysics* 49, 195.

Wolf, S. et al. (2012) Circumstellar disks and planets, science cases for the next-generation optical/infrared long-baseline interferometers, *Astronomy & Astrophysics Reviews* 20, 52.

Andrews, S.M. (2015) Observations of Solids in Protoplanetary Disks, *Publications of the Astronomical Society of the Pacific* 127, 961.

Sicilia-Aguilar, A. et al. (2016) A 'Rosetta Stone' for Protoplanetary Disks: The Synergy of Multi-Wavelength Observations, *Proceedings of the Astronomical Society of Australia* 33, 59.

Migration of planets:

Masset, F. & Snellgrove, M. (2001) Reversing type II migration: resonance trapping of a lighter giant protoplanet, *Monthly Notices of the Royal Astronomical Society* 320, L55, http://cdsads.u-strasbg.fr/abs/2001MNRAS.320L..55M.

Armitage, P.J. & Rice, W.K.M. (2005) Planetary Migration, *arXiv:astro-ph/0507492*

Kley, W. & Nelson, R.P. (2012) Planet-Disk Interaction and Orbital Evolution, *Annual Review of Astronomy & Astrophysics* 50, 211.

Chapter 7
Formation and evolution of planetary systems

The discovery of the variety of exoplanets has helped astronomers to better understand the transformation of the gas and dust disk of a proto-star into a mature planetary system, such as the Solar System: a construction which at first takes place rapidly with the formation of the nuclei of the giant planets as early as a few hundred thousand years, but which continues through the interactions between the giant planets and their parent disk and their migrations. A construction that is far from quiet, sometimes with the ejection of planets out of their system.

The formation of planets

The formation of planets in the protoplanetary disk of gas and dust is a well understood phenomenon, at least in broad outline, since the work of Viktor Safronov in the 1960s. The general idea is that the planets of the Solar System were formed by accumulation of solids of kilometric size, the planetoids, which themselves come from the agglomeration of smaller bodies, the planetesimals, formed by condensation of small grains. But this model faces a major problem: how to form giant planets before the gas and dust of the disk are dispersed by the radiation and the wind of the young star, in less than a million years? This question, as we shall see, has been answered only recently.

The initial dust grains may be of interstellar origin, or may originate from the condensation of the gas of the disk during its cooling. Closer to the star than the ice line, they contain very little water, but the grains farther away may be covered with ice. Low-velocity mutual collisions between the grains first form aggregates, then larger and larger solid bodies, the planetesimals, by gluing mechanisms that are not yet well understood. For the grains to grow into planetesimals, an important element is the fact that the gas and the grains do not revolve at the same speed around the star. The planets and planetoids, and all the bodies large enough, have a revolution dictated by the gravity of the star alone, like the planets in the Solar System: the so-called Keplerian revolution. On the other hand, the gas generally revolves at a slightly different velocity, for in the protoplanetary disk the density and therefore the pressure of the gas generally increase inwards, producing an outwardly directed force which adds to the centrifugal force corresponding to the Keplerian revolution. The small dust grains tend to follow this movement because they are strongly coupled with the gas by viscosity, and this has important consequences: the friction of the dust grains with the gas slows them down, causing their migration towards the interior of the disk and therefore towards the star. To form the planetesimals before the disk is dissipated, it is necessary that the grains stick to each other sufficiently rapidly to form larger bodies of decimetric size which are less coupled with the gas and will not fall on the star.

The problem of planetesimal formation has long appeared insurmountable, but a solution has recently been proposed, based on an idea put forward in 1972 by Fred Whipple (1906–2004): if there is a concentration of gas and dust, for example a dense ring such as we saw in the previous chapter, the dust will not be able to escape because the pressure of the gas decreases this time towards the interior, whereas the dust that drifts from the outside of the disc will accumulate there. In addition, large dust grains tend to drive the gas in their revolution, pushing it from the inside to the dense area and further stabilizing the system: this is a feedback. A dust trap has thus formed. This can even occur spontaneously, as shown in Figure 7.1. The dust grains or aggregates, which are now in a high-density region, can then rapidly agglomerate to form the planetesimals.

When the planetesimals reach a kilometric size, becoming planetoids, their gravitational attraction dominate the electromagnetic forces and the forces of entrainment by the gas, so that they continue to agglomerate by mutual collisions. This mechanism tends to accelerate, for the large planetoids attract the others more easily. However, this agglomeration has long seemed too slow to explain the formation, in less than a million years, of the nuclei of the giant planets, which reach ten solar masses in regions of the disk far from the star. This problem has recently been solved by showing that a 500–1000 km planetoid can grow by accreting solids a few centimeters in diameter – pebbles formed from grains: radio observations of the protoplanetary disks show that these pebbles are abundant at large distances from the star. The planetoid revolves around the star with a Keplerian velocity, while the gas rotates faster or slower if there is a pressure gradient, as we have seen above. The pebbles are partially driven by the gas and thus pass near the planetoid at low speed; they are then deflected by its gravity field and end up being captured. The small grains are driven at too high a speed to be accreted, while the largest pebbles are not driven by the gas and are in

Keplerian revolution: they are therefore more or less fixed with respect to the planetary seed. So only intermediate-size pebbles are captured by the planetoid. Numerical simulations show that the growth of the planetoid by this process is much faster than if it had formed by encounter and agglomeration of several large solids, all of which would have a Keplerian rotation.

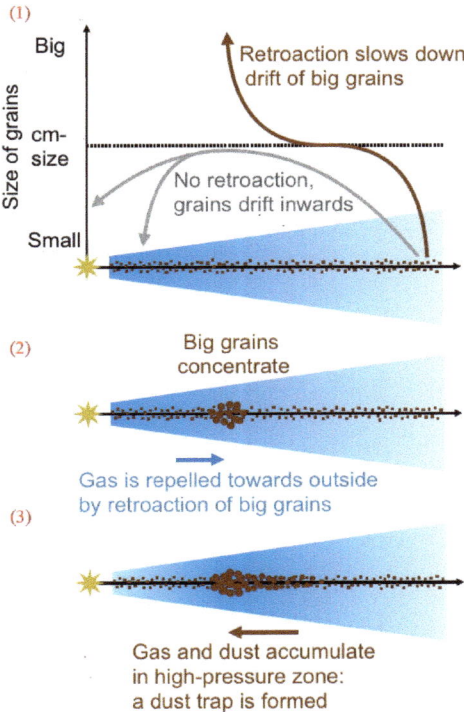

FIGURE 7.1 – Spontaneous formation of a dust trap, where the grains can grow to become planetesimals. The disk is schematized as seen edge-on, the gas is in blue and the dust in brown. In (1), the radial movement of the grains, slowed down for the larger grains by their feedback on the gas. In (2), the large grains concentrate accordingly, while the gas is pushed outwards under the effect of the feedback. In (3) the trap is formed and captures the gas and external dust; it has the form of a dense ring centered on the star. According to Gonzalez J.-F. et al. (2017) *Monthly Notices of the Royal Astronomical Society* 467, 1984.

The protoplanets grow in this way until the surrounding area subjected to their gravity is somehow emptied of its solid materials, and until they are dynamically isolated from one another. A fraction of the planetoids fragment as the result of their mutual collisions. In the Solar System, some of the resulting fragments end up on the Earth as meteorites when they cross our planet's orbit. The existence of two classes of meteorites, metallic and rocky, suggests that fusion and differentiation occurred in certain planetoids, inside which the heavy elements like iron fell to the center of the object, as was the case later in the planets. The heat that would have produced this fusion would have come from the decomposition of the ^{26}Al isotope of aluminum generated during the explosion of a nearby supernova that contaminated the initial material and possibly even triggered the collapse of the proto-solar nebula. ^{26}Al has a half-life of 700,000 years, and was therefore still very active at the time of planetoid formation. Another fraction of the planetoids escaped incorporation into planets and formed the asteroids and comets.

The protoplanets located inside the ice line contain only refractory elements, and the surrounding gas is too hot to be captured, or evaporate through heating by the radiation of the proto-star: they become terrestrial (also called telluric) planets. On the other hand, more distant protoplanets contain ice and when their mass is of the order of 10 Earth masses, they can gravitationally capture the surrounding gas, essentially made of hydrogen and helium, thus becoming giant planets. Most of the mass of Jupiter and Saturn is composed of this gas, but it makes only about 10% of the mass of Uranus and Neptune. Of course, capture of the gas by the protoplanetary core must take place before the gas from the disk has evaporated, and this is probably what limits the growth of the giant planets. The observation of exoplanets around very young stars of known age shows that Jupiters can actually form in less than a million years.

The evolution of planetary systems: what does the Solar System teach us?

Obviously, the Solar System is much better known than the exoplanetary systems, and it can teach us a lot about these systems in general. But it is the result of an evolution that lasted for 4.6 billion years. Knowing this evolution is a difficult problem, which was, however, addressed at the beginning of this century by scientists initially gathered at the Nice Observatory, hence the name given to their theory: the *Nice model* or *Nice scenario*. Logically, they first dealt with relatively recent periods. The older periods were the subject of further studies, involving many researchers who had worked on the Nice model, but also many new ones. Here, we present the model in the chronological order of the events.

It is assumed that Jupiter, which determined most of the subsequent evolution due to its considerable mass, was rapidly formed close to the water ice line, at a distance to the Sun of 3.5 a.u. (Figure 7.2). Closer to the Sun, there were only rocky planetsimals and some gas. Further away, there were gas and icy bodies of various sizes. It is in the intermediate zone, just beyond the ice line, that the amount of available material was maximum, and it makes sense to give birth to Jupiter there. The mass of the planet rapidly exceeded about a hundred terrestrial masses, resulting in a Type II migration into the Solar System. Saturn formed a little further, at a distance to the Sun of 4.6 a.u., but its mass was not yet sufficient to trigger a Type II migration. Jupiter quite quickly reached the orbit of Mars at 1.5 a.u., which had the effect of stopping the growth of this planet, assuming that it was already in formation.

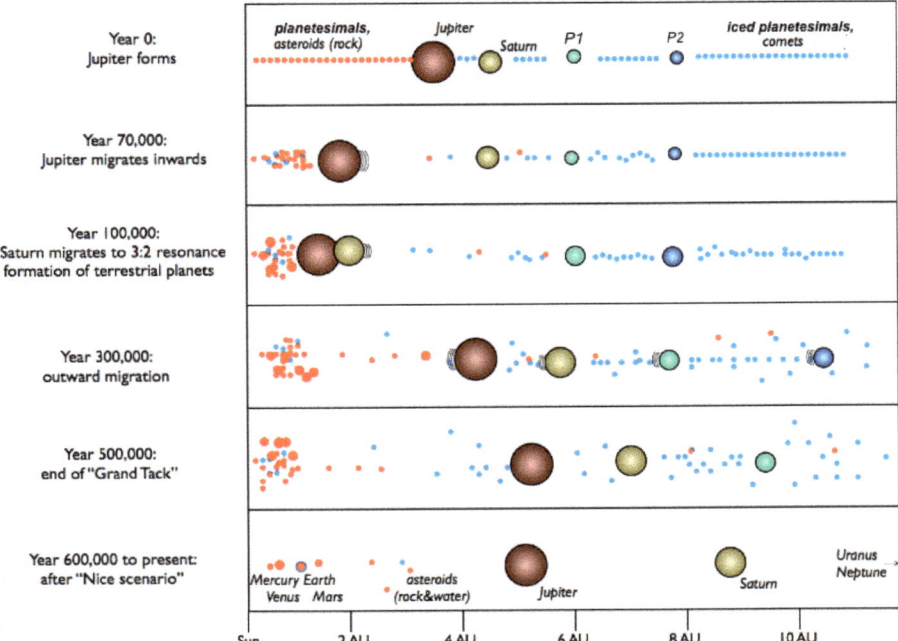

FIGURE 7.2 – The main stages of the first part of the initial Nice scenario. The planets are symbolized by circles whose size depends on their mass during the different stages. *P1* and *P2* will become Neptune and Uranus respectively (see text). The gas distribution of the protoplanetary disc is not shown. The planetesimals are schematized by red dots if they are purely rocky or blue if they are icy. Notice that the action of Jupiter pushes a fraction of the icy planetesimals into the inner regions of the Solar System. The large red dots symbolize the fusion of planetesimals into planetoids to form planets. The distance from these elements to the Sun is indicated by the bottom scale. The simulation stops 600,000 years after the formation of Jupiter. http://www.exoclimes.com/news/recent-results/a-scenario-for-the-formation-of-the-solar-system-the-grand-tack/.

Meanwhile, Saturn, still at 4.6 a.u., continued to grow (Figure 7.3). Its mass became finally large enough to trigger a Type II migration, which brought it rapidly inward, until the planet entered a 3:2 resonance with Jupiter, 3 periods of revolution of Jupiter being then equal to 2 periods of Saturn. The two planets, now digging a common gap in the disk, then began a slow outward migration while remaining blocked in 3:2 resonance, thanks to a mechanism that we explained in the previous chapter. This is what the authors of the model refer to as the "Grand Tack" in the language of seamen, which caused the two giant planets to reverse their motion like two racing sailboats revolving around a tag. It should be noted that the speed of their migration depended on the properties of the disk, in particular on its thickness, and might have been very slow.

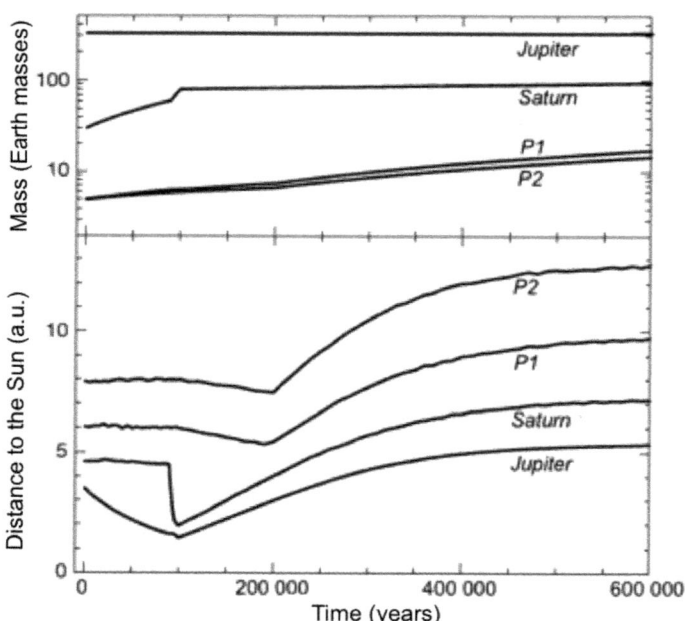

FIGURE 7.3 – Evolution over time of the mass of the four giant planets (top) and their distance to the Sun (below) in the initial Nice model. *P1* and *P2* were to become Neptune and Uranus respectively (see text). Time, on the x-axis, is counted from the end of the formation of Jupiter. At that time, the other three planets were still forming and gradually grew by capturing the surrounding matter. At 100,000 years, Saturn entered into 3:2 resonance with Jupiter. According to Walsh et al. (2011).

What about the other two giant planets, Uranus and Neptune? For now, we will call them *P1* and *P2* because we do not yet know which one became Uranus and which became Neptune. They formed around 6 and 8 a.u. respectively from the Sun. Their masses were insufficient to cause their migration. But when Jupiter and Saturn moved away from the Sun, *P1* and *P2* entered in resonance with each other (4:3 resonance), and with Saturn for *P1* (3:2 resonance, Saturn performing three revolutions when *P1* performed two revolutions). They were trapped and carried along in the same motion towards the outside. This configuration lasted until the gas dissipation of the protoplanetary disk was almost complete. The four planets were then fully formed, but a problem remains: if *P1*, which was formed closer to the Sun than *P2*, in a region where the quantity of material available was greater, was Uranus, it should have had a larger mass than *P2* (= Neptune?), while Neptune is actually more massive than Uranus. A mystery which, as we shall see, will be solved by advancing a little more in the history of the Solar System.

Indeed, we stopped in the evolution of the Solar System 600,000 years after the formation of Jupiter. But further upheavals were to come. There was little gas left, many planetesimals had gathered into planetoids and planets that were now fully formed, including perhaps the inner rocky planets (we will come back to this). An important disk of debris remained beyond the orbit of the planets, to about 30 astronomical units of the Sun. In the initial Nice model, its mass is assumed to be 35 times that of the Earth. This debris disk was composed of iced bodies: ice blocks, dwarf planets and cometary nuclei. The giant planets were on circular orbits in a more compact configuration than they are today, and they would still migrate. Jupiter was able to eject the fragments that are nearby: it losed energy in this way and slowly approached the Sun. Saturn and the other two massive planets were not able to do so, and instead gained angular momentum to the detriment of the debris, which laid mainly beyond their orbit: they slowly moved away from the Sun, the 3:2 resonance of Saturn with Jupiter being broken.

This evolution is shown in Figures 7.4 and 7.5. At first, everything was calm, but a total upheaval occurred when Saturn and Jupiter, moving away from each other, arrived in 2:1 resonance, the period of Saturn's revolution being then twice that of Jupiter. This occurred about 880 million years after the formation of the Sun. The orbits of all the planets were then deeply modified and became eccentric, notably that of Neptune, which was now farther from the Sun than Uranus. This solves the problem we had earlier with *P1* and *P2*. The planetoid disk was completely dispersed: part of it was in the form of icy objects outside the planets, forming the Kuiper belt and a distant comet reservoir, the Oort cloud. Another part bombarded the planets and their satellites: this is what is called the *Great Late Bombardment*. A major fraction of the craters on the Moon, Mercury and asteroids are the result of these impacts. There are hardly any more such craters on Venus, the Earth and Mars, because they have since disappeared by erosion, volcanism or plate tectonics. But it is thought that much of the water of the terrestrial oceans and water on Venus and Mars was brought by these fragments. Still today, comets are bringing water to the stratospheres of the four giant planets. Thereafter, the orbits of the planets were circularized under the gravitational effect of the debris disk. Since that time, in the 4 billion years that followed, the configuration of the Solar System has undergone only a few changes.

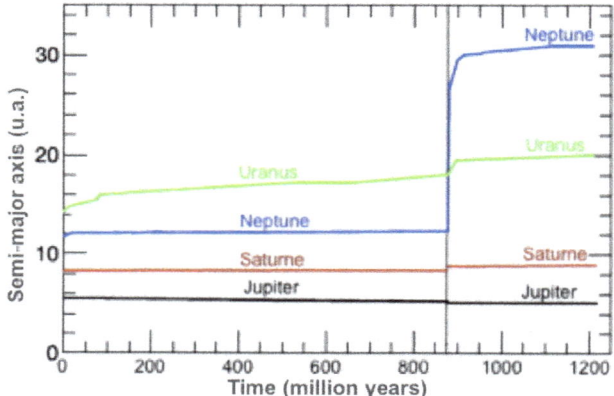

FIGURE 7.4 – The evolution of the orbits of the giant planets after their formation simulated numerically in the initial model of Nice. Under the gravitational effect of the outer debris disk, the planets migrated slowly: Jupiter approached the Sun slightly, while Saturn, Neptune and especially Uranus were moving away from it. Uranus was for the moment, and since its formation, farther from the Sun than Neptune. At the time when the period of Saturn's revolution became exactly twice that of Jupiter (2:1 resonance), a great upheaval occurred: the orbits were strongly altered and temporarily became very eccentric. Finally, the orbit of Neptune became external to that of Uranus. The orbits of the four giant planets no longer underwent any significant modification thereafter. According to Gomes, R. et al. (2005).

Formation and evolution of planetary systems

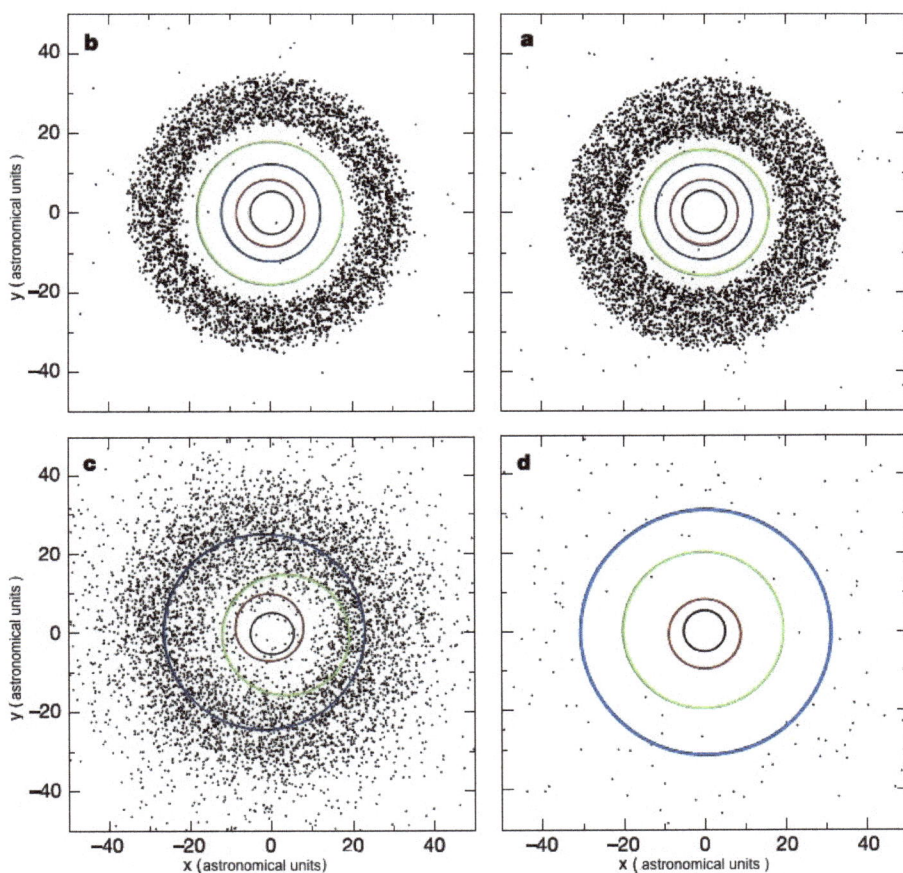

FIGURE 7.5 – The evolution of the solar system after the formation of the large planets, simulated numerically with the same parameters as for Figure 7.4. The orbits of the giant planets are represented: that of Jupiter in gray, of Saturn in red, of Uranus in green and of Neptune in blue. The dots represent what remained of the debris of the primitive Solar System after the formation of the planets, forming a ring beyond the orbit of the giant planets. **a**: 100 million years after their formation, the large planets started their migration; **b**: 879 million years after their formation, the orbits have changed and a 2:1 resonance between the periods of Jupiter and Saturn revolution occurred; **c**: 882 million years after formation, this resonance affected all the Solar System, significantly altering the orbits (Neptune's orbit was now outside Uranus's orbit) and precipitating the debris inwards and outwards, producing the Great Late Bombardment of the planets and the Moon; **d**: 200 million years later, the planets were in their quasi-definitive orbits and only 3% of the debris disk remained, forming in particular the Kuiper belt of icy objects. According to Gomes, R. et al. (2005).

Why no super-Earths in the Solar System?

The Nice model and its extensions, which represent one of the triumphs of contemporary celestial mechanics, succeeded to account for most of the characteristics of the Solar System, including the distribution and properties of asteroids, comets and icy objects, which we mentioned only in passing. It is true, of course, that various arbitrary assumptions had to be made in order to fit the results to observations, and the Nice model has different problems in its original form. Fortunately, it is possible to change the initial conditions and still arrive at a reasonable agreement with the observations. For example, in some simulations, the instability that produced the Great Late Bombardment occurred at the 5:3 resonance between Saturn and Jupiter, not at the 2:1 resonance. Much remains to be done, however, in order to account fully for the evolution of the Solar System, even if the principles are well understood.

However, the differences between the Solar System and what we know about exoplanetary systems are so large that they lead to other questions. For example, why does the Solar System not contain super-Earths of about ten Earth masses, which are so common among exoplanets? Astronomers generally believe that these super-Earths are the first planets to form in protoplanetary disks. However, while the planetesimals clustered very rapidly, perhaps in 1 million years, cosmo-chemical arguments suggest that the terrestrial planets might have been born only after 100 million years, well after Jupiter and Saturn. This seems somewhat paradoxical.

In order to solve this paradox, it has recently been suggested that super-Earths have indeed been able to form very early and have migrated until they were very close to the proto-Sun, as can be seen in nearly half of the exoplanetary systems. They would then have been destroyed, for example by falling on the proto-Sun, and the terrestrial planets would have formed afterwards. However, the destruction mechanism is very controversial.

Expelled planets, isolated exoplanets

During one of the above-mentioned violent episodes, it is possible that one or more planets have been expelled at large distances in the Solar System, or even out of it. This may have some relation with recent observations of the orbits of the icy objects of the Kuiper belt, which strongly suggest that they are perturbed by a ninth distant planet of about ten Earth masses occupying a very elliptic orbit (Figure 7.6).

Formation and evolution of planetary systems

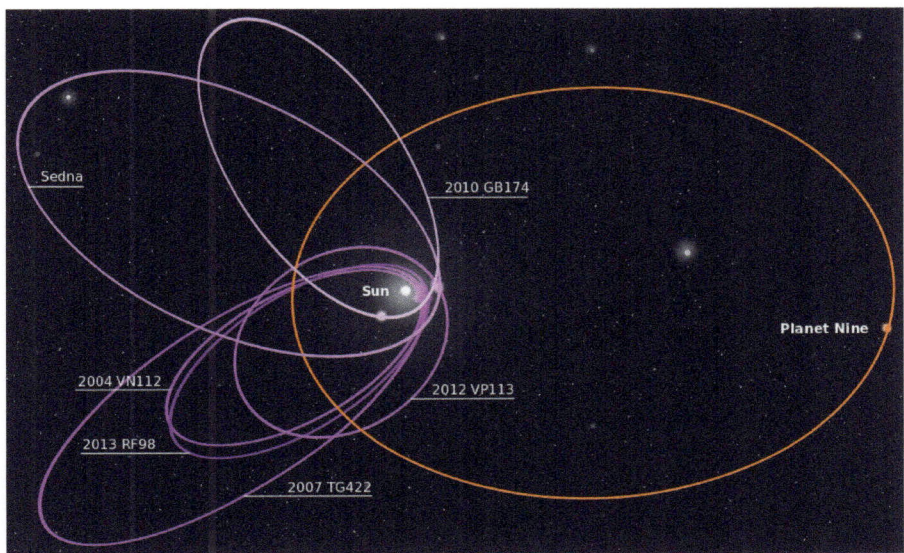

FIGURE 7.6 – Orbits of objects of the Kuiper belt with a half-axis of more than 250 a.u., seen in projection on the plane of the ecliptic. These orbits are not randomly oriented, but are all on the same side. Moreover, their perihelions are all approximately in the plane of the ecliptic and the objects all cross the ecliptic in the same direction, from the south to the north. The approximate orbit of the ninth hypothetical planet is indicated in orange. It would extend from about 280 to 1120 a.u. and would be inclined by 30° on the ecliptic. Wikimedia Commons.

Indeed, the initial mass of the Kuiper Belt was estimated to be about 30 Earth masses, in order to allow 100 km-size objects to form within a reasonable time, while the current mass is 100 times smaller. Most of the objects now missing from the Kuiper belt were ejected under the influence of Neptune, which gravitates at the inner limit of the belt; others have acquired the very large and eccentric orbits that we observe for some of them. These orbits were initially randomly distributed, and another disturbance was necessary to select the objects we see today, ejecting the other ones.

The most attractive explanation for this anomaly is the presence of a planet. It is possible to account for it with a planet of 10 terrestrial masses circulating on a very eccentric orbit, with a perihelion at about 280 a.u. of the Sun – ten times as far as Neptune – and an aphelion at about 1120 a.u. This would be the ninth planet of the Solar System. Unfortunately, the gravitational action of this planet being very slow, it is impossible to know from the theory where it is currently in its orbit. Indeed, a relatively simple analytical solution can be obtained by distributing the mass of the planet along its orbit. However, it has been possible to restrict its possible locations by examining the possible disturbances it may have on the motion of Saturn, which is known with great precision thanks to the telemetry of the Cassini probe, which orbited around this planet between 2004 and 2017. Since no effect can be detected, the planet cannot be close to its perihelion, and must be very distant: it will therefore be very difficult to discover.

Where can this planet come from? It certainly was not formed on the spot because there would not have been enough material available in the disk. Conversely, it could have formed between 5 and 20 a.u., and then been very strongly disturbed by Jupiter or by Saturn. However, the probability that it remained in the Solar System was low, as it would have been much more likely to be ejected. It could also have been ejected by another planetary system and captured by the Solar System. If we discover it someday, perhaps we shall know more about its origin.

The planets ejected by the exoplanetary systems are necessarily isolated in the interstellar medium. They could be extremely numerous, perhaps even more numerous than the ordinary stars, but their detection is difficult. Among the very few known isolated planets, the most interesting one, CFBDSIR J214947.2-040308.9, was discovered by a French-Canadian team with the 3.6-meter diameter CFH telescope in Hawaii, during a systematic search for very cold objects in the infrared. It is at a distance of 75 light-years and has a mass of a dozen Jovian masses. It could therefore be an intermediate object between a brown dwarf star and a very large planet.

What future for the Solar System?

The Solar System is 4.6 billion years old and at first glance looks quite stable; at least, this is what was thought since Laplace, though with some doubts. Jupiter and Saturn are now in the 5:2 resonance, which stabilizes them, and the mass of what remains of the disk of debris – asteroids and objects of the Kuiper belt – is far too weak to affect their motion, as well as that of Uranus and Neptune.

However, it is clear that there must have been changes, at least for the terrestrial planets. For example, Mars experienced a very different climate from today's one in a relatively recent past. Its climatic changes are generally attributed to large variations in the obliquity of the axis of the planet with respect to its orbital plane, the amplitude of which can reach 60° with a period of about 100,000 years. But there also could have been longer-term variations in its distance from the Sun and the inclination and eccentricity of its orbit.

This is what numerical simulations of the long-term evolution of the Solar System suggest. Planetary systems go through unstable and chaotic situations because tiny gravitational perturbations can produce dramatic changes. Although celestial mechanics can accurately predict the position of the terrestrial planets on a scale of tens of millions of years, their motion becomes chaotic on a longer scale and it is impossible to predict how the inner regions of the Solar System will look in a few hundred million years, and also to say what they looked like in the distant past (Figure 7.7). The effects are exacerbated if there is a resonance between the period of revolution of a terrestrial planet and that of Jupiter. This could happen to Mercury, and lead to its ejection. Collisions between planets are even possible in the distant future, albeit with very low probability.

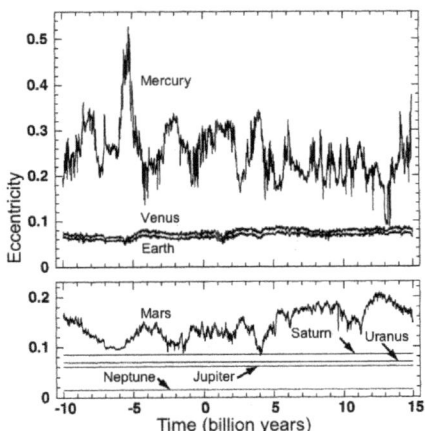
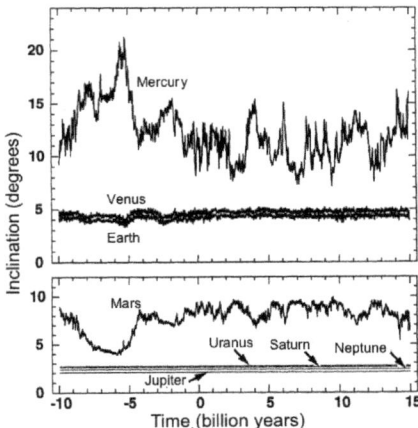

FIGURE 7.7 – An example of a simulation of the evolution of the eccentricity (left) and the inclination (right, the origin of the ordinates being arbitrary) of the orbits of the planets in the past and in the future. The semi-major axis of the orbits varies little, and the orbits of the giant planets are little affected by chaos. This is less the case for the large terrestrial planets, Venus and the Earth, and the behavior of Mercury and Mars is extremely chaotic and totally unpredictable in the long term (the fluctuations are not noise!). According to Laskar, J. (1994).

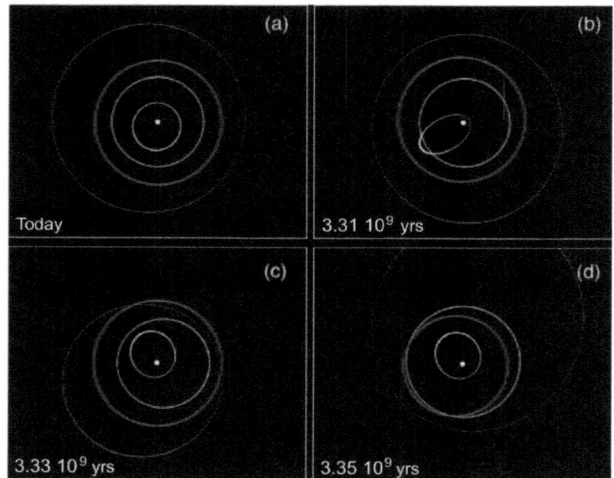

FIGURE 7.8 – Some possible future orbits for Mercury (white), Venus (green), the Earth (blue) and Mars (red). They result from numerical simulations made by Jacques Laskar and Mickael Gastineau. Time is expressed starting from the present time. (a) Present orbits. (b) A case where the Mercury orbit is sufficiently deformed after 3.31 billion years to allow a collision with Venus or with the Sun. (c) The orbit of Mars crosses the Earth orbit after 3.33 billion years. (d) This leads shortly after to a destabilization of the orbits which can also produce a close interaction or a collision between Venus and the Earth. (Thanks to J. Laskar and M. Gastineau)

What consequences for our understanding of exoplanetary systems?

The example of the Solar System has shown to us that the formation and evolution of any planetary system is an extremely complex affair, which depends greatly on the details of the configuration and the physical properties of the protoplanetary disk and even involves chaos. We are still far from understanding in detail how exoplanetary systems were formed and why they are so different from each other.

An important question concerns the large number of exoplanets with very eccentric orbits (whereas the planets of the Solar System have little eccentricity). When they are formed in the protoplanetary disk, the planets have approximately circular orbits. Later events are required to make certain orbits very elliptical. The theory and numerical simulations show that large eccentricities are the result of gravitational instabilities which almost necessarily occur when giant planets pass close to one another. In extrasolar systems, this seems to be the norm rather than the exception. Conversely, in the Solar System, the two biggest planets, Jupiter and Saturn, have never been close to one another; this is a fortunate coincidence as this rapprochement might well have occurred and the Earth as we know it would probably not exist. Uranus and Neptune, however, had eccentric orbits at the time of the Great Late Bombardment (see Figures 7.3 and 7.4). There was at that time an important disk of debris beyond these planets, whose gravitational action rapidly circularized the orbits. On the other hand, in systems where the residual debris disk has a low mass, the eccentric orbits persist.

The eccentricity of the orbits of giant planets profoundly affects the formation and dynamic behavior of terrestrial planets, which can then acquire very eccentric orbits themselves, to the point of falling on the star or being ejected. Simulations suggest that in such cases a single terrestrial planet may possibly subsist but with such an eccentric orbit that its habitability is quite problematic because of the enormous variations in its temperature. It is probable that habitable planets can only be found in systems devoid of giant planets, which could be the case around 90% of stars, or, if there are giant planets, in cases where the instability was relatively moderate, as in the Solar System.

We have seen that intermediate mass planets, say with less than 30 Earth masses, are frequent and usually have periods of revolution shorter than 100 days: they are so close to the star that they must be rather hot. The larger ones probably formed at larger distances from the star, as evidenced by their low density which indicates a large proportion of ice and gas, and then migrated inwardly as the remainder of the protoplanetary disk evolved. If this is true, the only reason why Uranus and Neptune are far from the Sun is that they were blocked by their resonances with Jupiter and Saturn, which dragged them outward during the final phase of the disk's evolution. The migration of Neptunes and super-Earths towards the star certainly affects the formation of planets of mass close to that of the Earth. But the super-Earths themselves could be habitable if they are at the proper distance from the star. So, habitable planets are not necessarily rare, but they are probably most often bigger than the Earth: they have to be taken into account in our search for habitable planets.

Thus, it may be thought that planets similar to the Earth must be rather infrequent, and that the habitable planets are mostly objects of larger mass, some of which could be nuclei of giant planets that have failed to capture much gas for some reason, and which migrated inward into the habitable zone. Being formed beyond the ice line, they should contain a lot of water and therefore be favorable to life.

Finally, it should be noted that the interactions between two protostellar systems close to one another, or between such a system and a nearby star, as this must be frequent since the stars are often binary or tend to form in groups, can cause profound changes in the planetary orbits that could become very eccentric, or even eject the planets. This could also favor the formation of hot Jupiters: indeed, one seems to find more hot Jupiters in stars belonging to clusters than around isolated stars.

The formation and evolution of planetary systems is therefore an extremely complex problem. The structure of the protoplanetary disks is governed by several physical phenomena not always well understood, such as the turbulent viscosity of the disk, which determine in particular the position of the ice line, and therefore the formation and migration of the planets. The physics of planetesimal formation and growth is also poorly understood. As for the migration of the planets, we have seen its fundamental role for the evolution of the Solar System. But the theory predicts that the migration of giant planets is so rapid that one wonders why there are not more hot Jupiters, while most Jupiters are at more than 1 a.u. from their star (see Figure 5.9). Is there a mechanism to slow down their migration? Or, would the giant planets that were formed sooner have been swallowed by the star, while those that we see today would have formed later, without having time to reach the inner regions before dissipation of the disk? Finally, are the terrestrial planets and super-Earths of the same nature?

There remains a lot of work to be done to understand the exoplanetary systems, both in terms of theory and numerical simulation, and in terms of observation. It is essential to observe not only the planets themselves but also the protoplanetary disks, the transition disks and the debris disks in exoplanetary systems. The latter disks must be frequent, since at least half of these systems show an infrared excess produced by the thermal emission of circumstellar dust; also, the Spitzer satellite observations have shown that at least 20% of solar-type stars possess debris disks, often very extended. It will also be important to identify cases where resonances exist between the exoplanets of the same system, as they can give us some indication of the origin and evolution of this system.

Bibliography

Given the complexity of the subject, it is more extensive than in the other chapters.

Mechanisms of planetary formation:
Lissauer, J.J. (1993) Planet Formation, *Annual Review of Astronomy & Astrophysics* 31, 129, http://cdsads.u-strasbg.fr/abs/1993ARA%26A..31..129L.

Pollack, J.B. et al. (1996) Formation of the Giant Planets by Concurrent Accretion of Solids and Gas, *Icarus* 124, 62.

Blum, J. & Wurm, G. (2008) The Growth Mechanisms of Macroscopic Bodies in Protoplanetary Disks, *Annual Review of Astronomy & Astrophysics* 46, 21.

Lambrechts, M. & Johansen, A. (2012) Rapid growth of gas-giant cores by pebble accretion, *Astronomy & Astrophysics* 544, A32, http://cdsads.u-strasbg.fr/abs/2012A%26A...544A..32L.

Gonzalez, J.-F., Laibe G. & Maddison S.T. (2017) Self-induced dust traps: overcoming planet formation barriers, *Monthly Notices of the Royal Astronomical Society* 467, 1984.

The Nice model and its developments:
Tsiganis et al. (2005) Origin of the orbital architecture of the giant planets of the Solar System, *Nature* 435, 459.
Gomes et al. (2005) Origin of the cataclysmic Late Heavy Bombardment period of the terrestrial planets, *Nature* 435, 466.
Morbidelli, A. et al. (2007) Dynamics of the giant planets of the Solar System in the gaseous protoplanetary disk and their relationship to the current orbital architecture, *Astronomical Journal* 134, 1790, http://cdsads.u-strasbg.fr/abs/2007AJ....134.1790M.
Walsh, K.J. et al. (2011) A low mass for Mars from Jupiter's early gas-driven migration, *Nature*, 475, 206.
Raymond, S.N. & Morbidelli, A. (2014) The Grand Tack model: a critical review, in *Complex Planetary Systems, IAU Symposium 310*, 194, http://cdsads.u-strasbg.fr/abs/2014IAUS..310..194R.
Batygin, K. & Laughlin, G. (2015) Jupiter's decisive role in the inner Solar System's early evolution, *Proceedings of the National Academy of Sciences of the USA* 112, 4214, http://www.pnas.org/content/112/14/4214.full.
Batygin, K. & Brown, M.E. (2016) Evidence for a distant giant planet in the Solar System, *Astronomical Journal* 151, 22.
Fienga, A. et al. (2016) Constraints on the location of a possible 9th planet derived from the Cassini data, *Astronomy & Astrophysics* 587, L8.

Formation and evolution of exoplanets:
Wolf, S. et al. (2012) Circumstellar disks and planets, science cases for the next-generation optical/infrared long-baseline interferometers, *Astronomy & Astrophysics Reviews* 20, 52.
Morbidelli, A. (2014) Scenarios of giant planet formation and evolution and their impact on the formation of habitable terrestrial planets, *Philosophical Transactions of the Royal Society A* 372, p. 2013.0072.
Morbidelli, A. & Raymond, S.N. (2016) Challenges in Planet Formation, *Journal of Geophysical Research: Planets*, 121, 1962
Shara, M. et al. (2016) Dynamical interactions make hot Jupiters in open star clusters, *Astrophysical Journal* 816, 59.

Chaos in the Solar System:
Laskar, J. (1994) Large-scale chaos in the solar system, *Astronomy & Astrophysics* 287, L9, http://cdsads.u-strasbg.fr/abs/1994A%26A...287L...9L.
Laskar, J. & Gastineau, M. (2009) Existence of collisional trajectories of Mercury, Mars and Venus with the Earth, *Nature* 459, 817.

Chapter 8
The physical nature of exoplanets

For most exoplanets, we know little more than their radius or a lower limit of their mass; with this only, it is difficult to launch into comparative planetology. But there is more detailed data for a few hundred objects, especially on their density; in addition, more and more observations give indications on the atmospheres of exoplanets. Not surprisingly, this growing body of information reveals worlds that are in some ways very different from the familiar planets of the Solar System.

The observables

Among the different methods for detecting exoplanets, two of them (velocimetry and detection by transit), when combined, allow to determine the basic characteristics of exoplanets. The transits give us the radius of the object, and velocimetry, in the case of a planet also observed by transit, gives a measurement of its mass. Having thus access to the density of exoplanets, we could hope, by analogy with the planets of the Solar System, to determine whether it is a giant or a rocky planet.

The reality, however, is more complex. Solar system planets fall into two distinct categories, the giants with a mass larger than 10 Earth masses, and the terrestrial ones whose mass is of the order of a terrestrial mass. This dichotomy can be simply explained by the nucleation formation scenario, which predicts that an object of mass greater than about 10 Earth masses has a gravity field large enough to accrete the surrounding protoplanetary gas, essentially composed of hydrogen and helium; the object then becomes a giant planet. However, the mass distribution of all the detected exoplanets (see Figure 5.1) does not reflect this dichotomy: a continuous range of masses is observed, from the Earth's mass to more than 1000 terrestrial masses. So we do not find at all the simple scenario we could have expected when considering the Solar System.

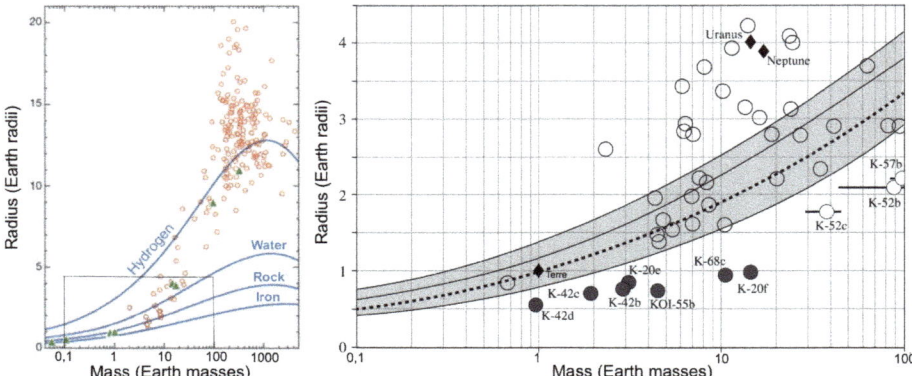

FIGURE 8.1 – Left: mass-radius diagram of the exoplanets observed by transit for radii smaller than 20 terrestrial radii; the blue curves show different models of internal composition predicted for these planets. The planets of the Solar System are represented by green triangles (from Howard, A.W. et al., 2013). Right: enlargement of the rectangle of the left diagram, with more data. The lower curve limiting the shaded region corresponds to the density of iron and the upper one corresponds to that of water; the dashed line is for the terrestrial composition (rock). The data for the planets represented by full gray circles are uncertain. The planets above the curves contain gaseous hydrogen and helium (from Mocquet, A. et al., 2014). It can be seen that existing models do not adequately account for the observed diversity of planets.

Figure 8.1 illustrates this situation: it is impossible to account for the variety of exoplanets with the two categories of rocky and giant planets only. In particular, objects of small size are seen, the density of which reaches that of silicates and even that of iron. Located very close to their star, they can be hot enough (over 2000 K) for their side facing the star to be in the form of magma. This is in particular the case with CoRoT-7 b, one of the first small planets detected by transit. But there is another category of objects even more astonishing, whose density exceeds that of iron! This is the case of Kepler-52 b, Kepler-52 c and Kepler-57 c, which is shown in the lower right corner of the right diagram of Figure 8.1. As suggested by a team from the Nantes Planetology Laboratory, these objects could be what remains of ancient giant planets, originally formed far from their stars, which would have migrated towards their stars, gradually getting rid of their gaseous envelope to finally retain only their core, which had been subjected to a high rate of compression in the first phase of formation of these objects.

Moreover, the other end of the domain of densities also offers us surprises. We have seen (Chapter 5, Figure 5.9) that there exists a large number of giant exoplanets with a mass comparable or superior to that of Jupiter, and whose radius is clearly greater than that expected for a planet formed essentially of hydrogen: they are called "inflated Jupiters". Located in the immediate vicinity of their star, these objects could be subject to significant atmospheric escape, as was suggested for the hot Jupiter HD 209458 b, the first exoplanet observed in transit.

How can we better define the nature of an exoplanet using the parameters we know? Another key parameter is the equilibrium temperature of the planet surface. This temperature obviously depends on the distance of the object to its host star, as well as on the spectral type of this star, which indicates whether it is more or less hot, with a maximum of radiation in the ultraviolet, blue or red spectral range according to its temperature. We can estimate the equilibrium temperature by making a hypothesis on the albedo of the planet (i.e. the percentage of the stellar flux that it reflects) and on its rotation velocity, which affects the redistribution of heat on the planet (see the box in Chapter 5, p. 72). For planets close to their stars, on orbits of a few tens of days, numerical models show that they can be assumed to be in synchronous rotation with their stars, as is the case with the Earth-Moon system and satellites close to the giant planets. This particular configuration has important consequences for all hot exoplanets, whether small or giant: since they always show the same hemisphere to their host-star, the atmospheric circulation must be very active at least in the case of giant planets, and the surface properties can be very different on the day and night sides in the case of rocky exoplanets. For the planets that are farther from their stars, their rotation is unknown, but we can then assume a rapid rotation, as in the case of the planets of the Solar System beyond the orbit of the Earth, so that the heat is well redistributed.

From the temperature of the exoplanet and its mass, it is possible to have a first estimate of its atmospheric composition using the reactions of thermochemical equilibrium. These reactions allow us to calculate, assuming that the chemical elements have the same abundance as in the Sun, what the chemical composition of a gas is going to be as a function of its temperature and its pressure. In particular, they predict that carbon and nitrogen evolve respectively as methane CH_4 and ammonia NH_3 at high pressure and low temperature and, under the opposite conditions, as carbon dioxide CO_2, carbon monoxide CO and molecular nitrogen N_2. Thus, the thermochemical equilibrium explains the presence of methane and ammonia in the giant planets of the Solar System and the presence of carbon monoxide (converted to a large extent to CO_2 by reaction with H_2O) and nitrogen in the atmosphere of the terrestrial planets. The temperature/mass diagram of the exoplanets thus allows a first classification of the exoplanets according to their chemical composition (Figure 8.2).

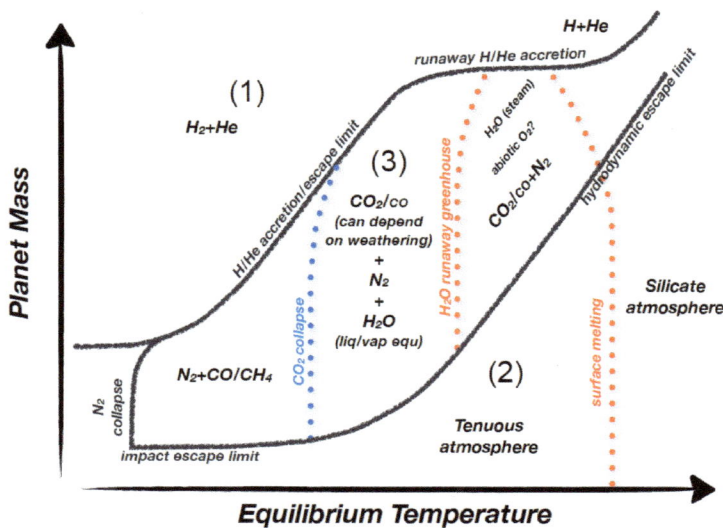

FIGURE 8.2 – Schematic diagram of the different classes of exoplanets, as a function of the temperature of their atmosphere (assumed to be in equilibrium) and their mass. The thick curves delimit transitions between different regimes. To the left of the upper curve, hydrogen and helium do not escape from the atmosphere, as in the relatively cold giant planets; to the right of this curve there is no more hydrogen and helium. Below and to the right of the bottom curve, the atmosphere is ejected more or less completely as a result of cometary or meteoritic impacts, or by evaporation due to heating by the radiation of the star. From Forget, F. & Leconte, J. (2014).

Figure 8.2 clearly shows three, and not two, types of planets (see also the box in Chapter 5). The region of high mass and low temperature is the domain of gas giants, composed mainly of hydrogen and helium. Depending on the fraction of gas they contain (which is undoubtedly related to their position in relation to the ice line at the time of their formation), they may be "gaseous giants" such as Jupiter and Saturn, or "icy giants", as Uranus and Neptune. Under the opposite conditions, one finds solid planets with a tenuous atmosphere or even without atmosphere, which can be very hot. None of these planets exists in the Solar System. The intermediate region, which contains Venus, the Earth and Mars, is where molecular nitrogen, carbon dioxide, some carbon monoxide, and also water, possibly in liquid form, can be found; the temperate part of this region is called the habitable zone.

The first measurements of the atmospheric composition of hot Jupiters

To find out more, there is only one solution: to directly determine the atmospheric composition of exoplanets. A method has emerged over the past ten years: spectroscopy, in the visible and infrared range, of exoplanets in transit in front or behind their stars. We have seen (Chapter 3) that the spectroscopy of these objects is possible in two different configurations: (1) during a primary transit, when the planet passes in front of the star, the spectrum of the planet can be obtained at the terminator (day/night transition), in transmission in front of the stellar spectrum; (2) during a secondary transit, just before and after the planet passes behind the star, we can observe the emission spectrum on the day side of the planet. In both cases, the light of the star before, during and after transit has to be measured very precisely, and a difference is made to subtract the light coming from the star alone. In the case of a primary transit, the signal is stronger if the planet is voluminous, not very dense and close to its star; in the case of a secondary transit, the signal is stronger for hot planets. In all cases, the planet/star flux contrast is higher for small and relatively cool stars, of type M and to a lesser extent K.

The first spectroscopic measurements in primary transit were carried out in the ultraviolet and visible domains in the early 2000s with the COS and STIS spectroscopes of the Hubble Space Telescope, and also from the ground in the visible range. Sodium and potassium atoms, as well as oxygen, carbon, and hydrogen escaping from the exosphere of hot Jupiters like HD 189733 b and HD 209458 b, have been detected; both exoplanets are relatively easy to observe because they are close to us. These measurements also revealed the presence of clouds and mists, attributed, given the temperature of the planet, to titanium (TiO) and vanadium (VO) oxides in solid form. A little later, the infrared range, better adapted to the detection of the neutral molecules, became accessible to the observations by transit, thanks to two very powerful instruments: the NICMOS-WFC3 spectrometer of the Hubble Space Telescope, in the near infrared, and the IRAC-IRS-MIPS spectrometer on the Spitzer satellite in the middle infrared around 10 µm. For example, the molecules of water, methane, ammonia, dioxide and/or carbon monoxide have been detected in more than 20 exoplanets, mainly hot Jupiters (Figure 8.3).

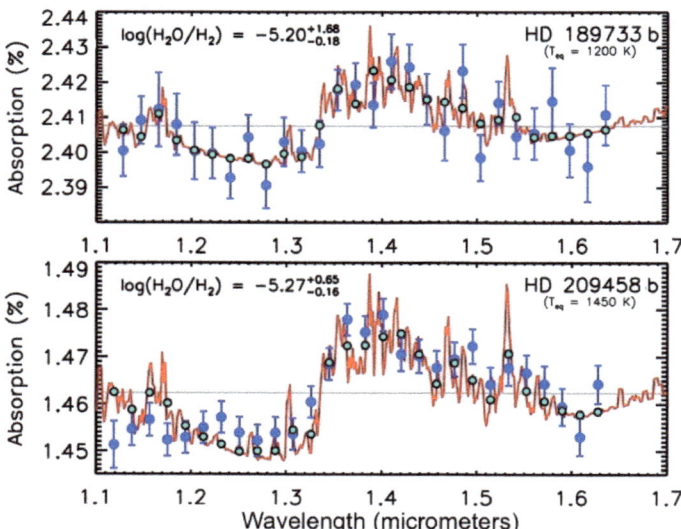

FIGURE 8.3 – The transmission spectra of two hot Jupiters, HD 189733b (top) and HD 209458b (bottom), obtained in the near infrared with the Hubble Space Telescope WFC3 instrument (see also Figures 3.6 and 3.7). The data are shown in blue and the models are in red. The increase in absorption around 1.4 µm corresponds to a band of H_2O. From Madhusudhan et al. (2014).

In parallel with primary transit spectroscopy, the measurements made at the time of the secondary transits provide important additional information. While the transmission measurements give the amount of molecules present on the line of sight at the terminator, between the star and the observer, the measurements in secondary transit provide a direct measurement of the planet's emission, either coming from the reflected stellar light or from the thermal emission of the planet. However, in almost all transit observations carried out to date, the planet is very close to the star and its equilibrium temperature is very high, often higher than 1000 K. In this case, the thermal emission of the planet is much larger than the reflected stellar radiation, even in the near infrared at wavelengths larger than 1 µm. The interpretation of the planetary spectrum is more complicated but brings more information than that of primary transit spectra. The secondary transit spectrum depends on the temperature of the emitting region (Figure 8.4); if the exoplanet has a stratosphere (where the temperature increases above a certain altitude), the molecular bands may appear in emission, possibly superimposed on bands formed in absorption deeper in the stratosphere.

The physical nature of exoplanets

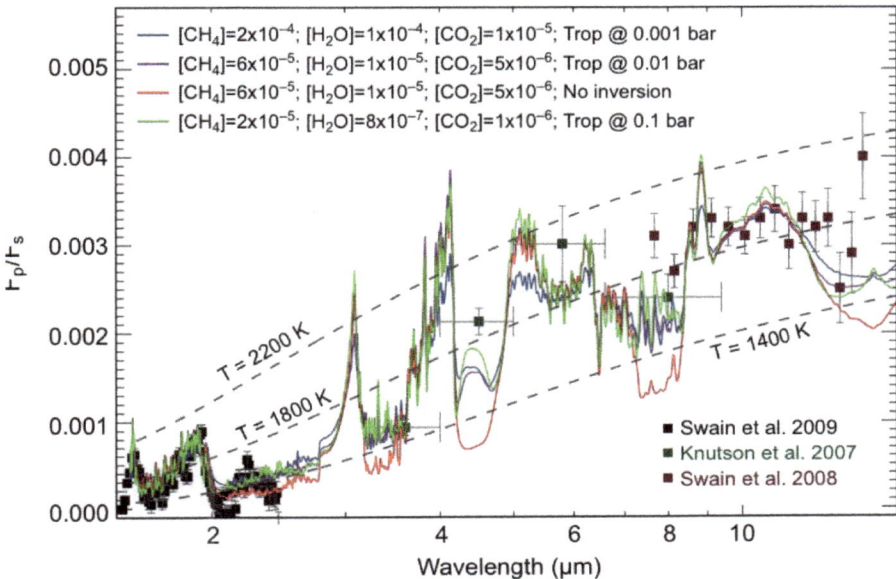

FIGURE 8.4 – Examples of emission spectra obtained on the exoplanet HD 209458 b, compared to different models (color curves) with or without a temperature inversion, thus having a tropopause below which the molecular bands are in absorption and above which they are in emission. A significant difference is observed between the observations and the predictions of the model without inversion at 4.3 µm (CO_2) and 7.7 µm (CH_4). The data suggest a temperature inversion in the atmosphere of the planet. However, this inversion is controversial. From Swain, M. R. et al. (2009).

Thus, in principle, the study of secondary transits makes it possible to obtain some information on the temperature profile in the atmosphere of the observed exoplanet. This work was done in the case of some bright objects, including the two hot Jupiters HD 189733 b and HD 209458 b mentioned above (Figure 8.5). Then the first difficulties arose. Indeed, the thermal profiles measured for the two hot Jupiters were in such a range that, according to the reactions of the thermochemical equilibrium, the carbon should have been in the form of CO and the nitrogen in the form of N_2; however, in both cases, methane CH_4 was detected in addition to CO and CO_2 (Figure 8.4). The reality is therefore more complex than anticipated by the photochemical equilibrium models.

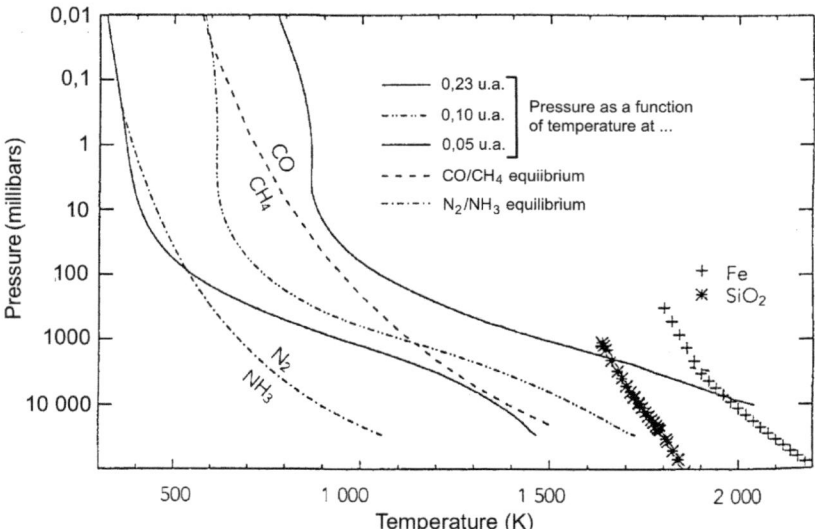

FIGURE 8.5 – Chemical structure of the atmosphere of a giant exoplanet located near a solar-type star, for three values of its distance to the star. The dotted curves indicate the expected changes, assuming thermochemical equilibrium, between N_2 and NH_3 and between CH_4 and CO, as a function of temperature and pressure. The condensation curves of iron and SiO_2 are also shown. It can be seen that, for hot Jupiters, an atmospheric composition dominated by CO and N_2 (apart from hydrogen and water) is expected, whereas CH_4 is actually detected on the two exoplanets for which the atmospheric composition is known: they are not in thermochemical equilibrium. From Encrenaz, T. et al., *The Solar System*, EDP-Sciences/CNRS-Éditions (2003).

Possible causes of departure from thermochemical equilibrium

Which mechanisms can explain the departure from thermochemical equilibrium observed on some hot Jupiters? As is often the case in the study of exoplanets, the study of the giant planets of the Solar System can bring us some information. In the case of Jupiter and its neighbors, we have identified two processes responsible for non-equilibrium chemistry: vertical gas transport (active in their deep troposphere) and photochemistry (which takes place in their stratosphere).

Let us consider first the vertical transport. These can lead to out-of-equilibrium chemistry if, in the deep troposphere, the vertical transport rate is greater than the chemical destruction time of some molecules. This is particularly the case for phosphine PH_3 in the deep troposphere of Jupiter and

Saturn. At a temperature of 800 K and a pressure of several hundred bars, thermochemical equilibrium predicts the reaction of PH_3 with H_2O to form P_4O_6. Phosphine should therefore not be present at a few bars, which is the pressure level probed by infrared measurements. Its presence on Jupiter and Saturn can only be explained by a very active vertical transport that carries phosphine upwards in a time shorter than the destruction time of this molecule. The same process is observed with CO in the troposphere of Neptune. A similar mechanism could explain the presence of CO in giant exoplanets.

As for photochemistry, it is the direct consequence of the irradiation of planetary or exoplanetary atmospheres by the ultraviolet radiation of their star. It is known that the photochemistry of methane, very active in the stratosphere of the giant planets of the Solar System, leads to the formation of hydrocarbons on the four giant planets, starting with ethane C_2H_6 and acetylene C_2H_2. Although the abundance of these two molecules is very low (some 10^{-6} at most as compared to hydrogen), their effect on the thermal spectrum is very important because of the high intensity of their spectral bands in the middle infrared. In the case of giant exoplanets close to their stars, one can expect a very important photochemistry linked to the high stellar irradiation (Figure 8.6).

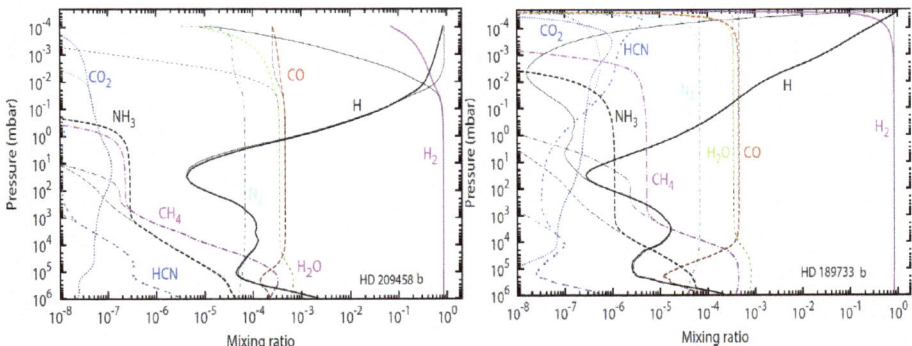

FIGURE 8.6 – Chemical structure of the atmosphere of two hot Jupiters, HD 209458 b (left) and HD 189733 b (right) calculated on the assumption of thermochemical equilibrium (black lines) and with inclusion of vertical transport and photochemistry (color lines). The difference in thermochemical equilibrium is less noticeable in the case of HD 209458 b because, since the temperature of the planet is higher, its structure is dominated by thermochemistry. From Vénot et al. (2012).

The observations indicate that, as for the giant planets of the Solar System, important vertical motions of gas, and chemical reactions initiated by the ultraviolet radiation of the star take place in the atmospheres of giant exoplanets.

Clouds and mists on exoplanets

In the planets of the Solar System, we currently observe the condensation of various ices, including the water ice on Earth and Mars, and solid methane and its photochemical derivatives in giant planets. The situation is quite different in exoplanets whose equilibrium temperature can in some cases exceed 2000 K. Assuming the same abundances of elements as in the Sun and using the reactions of thermochemical equilibrium, it is possible to determine the cloud structure of an atmosphere as a function of temperature. Thus, in the atmosphere of giant planets, the expected clouds are made of solid methane at 80 K, ammonia at 145 K, NH_4SH and NH_4OH at about 200 K and H_2O around 250 K. At greater depth, we can expect the condensation of potassium chloride and other chlorides, zinc sulfide and other sulfides at a few hundred K, then SiO_2, $LiCl$, Na_2S, Fe and magnesium silicates between 800 and 1500 K. Around 2000 K, the formation of corundum (Al_2O_3), as well as titanium (TiO) and vanadium (VO) oxides, is expected. The presence of these latter oxides has been suspected on some very hot planets on the basis of the shape of their visible spectrum; they were also detected on some brown dwarfs. Their presence could be at the origin of an opacity sufficient to create a temperature inversion. TiO has recently been detected in the transmission spectrum of the hot Jupiter WASP-19 b.

Spectroscopic measurements of super-Earths and exo-Neptunes in transit

In the wake of transit observations of hot Jupiters, several transit observations were made on the smaller objects; these were found to be possible when the host star was also of small radius. This is the case in particular with GJ 1214 b, a super-Earth in orbit close to a M star; its equilibrium temperature, of the order of 600 K, suggests that it could contain water vapor. The primary transit observations revealed a surprisingly flat spectrum (Figure 8.7). The lack of spectral signature can be explained by the presence of mist, possibly consisting of potassium chloride or zinc sulfide, maintained at high altitude by an intense vertical circulation.

The exoplanet GL 436 b is another example of a favorable target for transit spectroscopy, as this "hot exo-Neptune" of 22 Earth masses is also in orbit near a M star and has an equilibrium temperature of about 500 K. The presence of hydrogen as well as CH_4, CO_2, CO and H_2O was detected in this object. According to the ultraviolet and visible transit measurements, the exoplanet could be surrounded by a cloud of hydrogen that would escape like a cometary tail. The observed transmission spectrum led some authors to suggest a very high metallicity (i.e. a high hydrogen depletion), and heating due to a strong tidal effect, but without achieving a satisfactory solution (Figure 8.8).

The physical nature of exoplanets 117

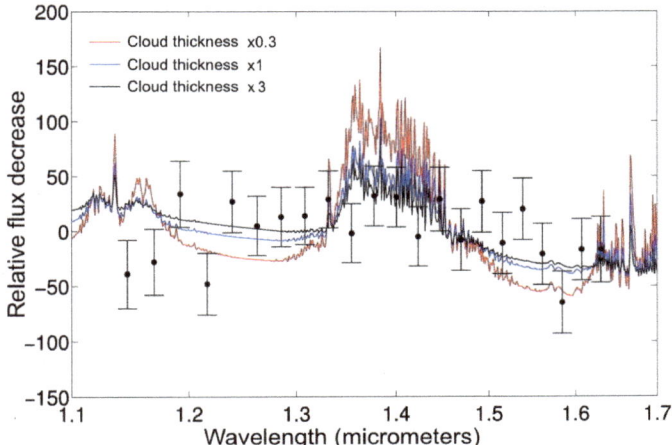

FIGURE 8.7 – The transmission spectrum of the super-Earth GJ 1214 b (black points with error bars) compared to different models corresponding to different abundances of clouds at high altitude. The spectral signatures expected around 1.14 and 1.38 µm are due to the water vapor, presumed to be the dominant species. Their absence in the observed spectrum can be explained by the presence of a cloud screen at high altitude. From Charnay, B. et al. (2015).

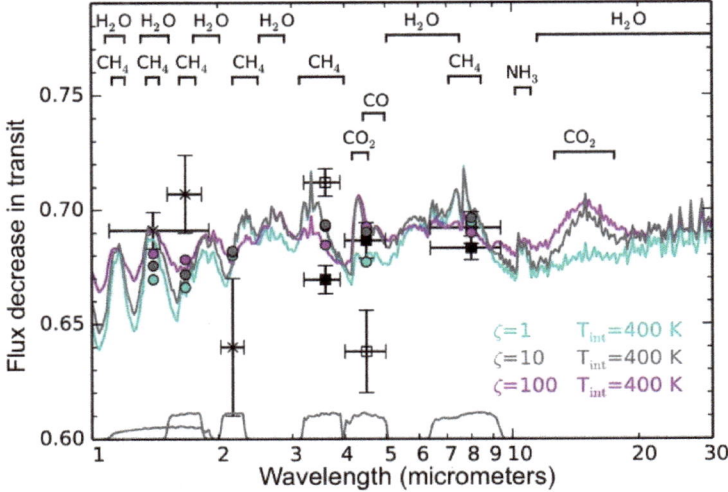

FIGURE 8.8 – The transmission spectrum of the hot exo-Neptune GJ 436 b, compared to different models involving different values of the metallicity ζ (i.e. the ratio of heavy elements compared to hydrogen, the solar value being equal to 1). The points correspond to different observations and the wavelength location of the bands of various molecules is indicated. It can be seen that no model can account for the measurements. From Agundez et al. (2014).

As in the case of hot Jupiters, the atmospheric composition measured in less massive exoplanets show deviations from the predictions of the photochemical equilibrium. Models have been developed to try to understand these differences, considering in particular the case of GJ 3470 b, another hot exo-Neptune in orbit around a M star, whose characteristics are intermediate between the two cases mentioned above (Figure 8.9). The results of these models show that in most cases methane and its photochemical derivatives are the dominant species. However, the CO/CH_4 ratio may become greater than unity in the case of a high metallicity and a temperature higher than 1500 K.

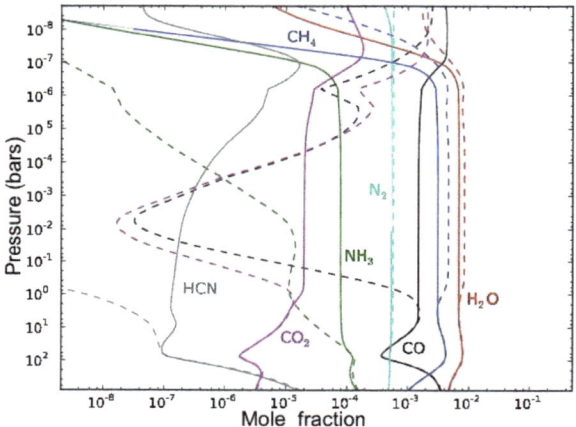

FIGURE 8.9 – Vertical distributions of molecular abundances in the case of the hot Neptune GJ 3470 b under the hypothesis of thermochemical equilibrium (interrupted lines) and including non-equilibrium processes (vertical transport, photodissociation, solid lines). It can be seen that these processes favor the presence of NH_3, CO and HCN. From Vénot et al. (2014).

Spectroscopy of exoplanets from the ground

A very original method, initiated by I. Snellen and his colleagues in Leiden, allowed the detection of the CO molecule from the ground in the near infrared, in the hot Jupiter HD 209458 b; we have seen that this target is favorable because its host star is particularly bright. The method consists in measuring the Doppler-Fizeau effect of the CO planetary lines due to its revolution around the star, using a very large telescope (in this case the VLT) and a very high spectral resolution power (100,000 with the CRIRES imaging spectrometer). The method also provides an estimate of the velocity of the atmospheric winds between the day and night sides of the planet. Later, the same method allowed the detection of CO and H_2O on other sources, including exoplanets without transit, in particular τ Boo b and 51 Peg b.

Finally, a new category of exoplanets can now be studied spectroscopically in the visible and near infrared range: these are the giant exoplanets detected by direct imaging. Thanks to the development of instruments specially designed for this research, in particular on the Keck, Gemini and VLT telescopes, it is now possible to obtain images of planets located at a few tens or hundreds of astronomical units of their star and to make their spectrophotometry at low resolution. Among the first successes of this research area is the four-planet system discovered around the young, massive star HR 8799 (see Appendix 2). The first three planets of this system, with about ten Jovian masses, were discovered by Christian Marois and his colleagues at the Keck and Gemini telescopes at distances ranging from 25 to 70 a.u. from their star; a fourth planet of the same size was then found at about 14 a.u. from the star. These relatively young planets still emit, by gravitational contraction, enough energy for their infrared radiation to be detectable. The spectrophotometric measurements carried out in the infrared bands observable from the ground made it possible to constrain the diameter and the temperature of these objects, using transfer models including clouds, derived from models of brown dwarfs (Figure 8.10). In the future, with better spectral resolution, it will become possible to identify the molecules present in the atmosphere of these objects of a new type.

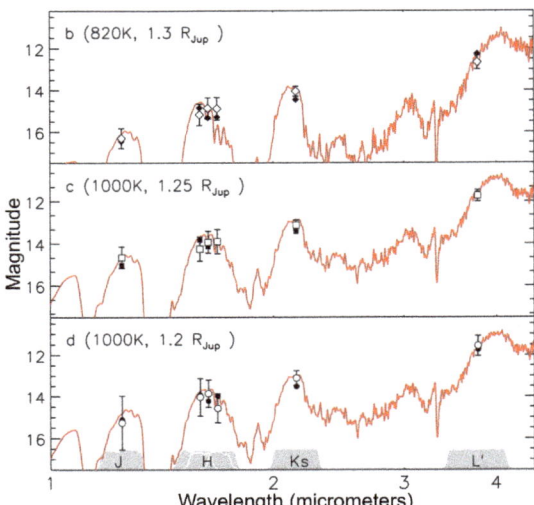

FIGURE 8.10 – Spectrophotometry of three exoplanets (b, c, d) discovered around HR 8799. The red curves show the synthetic spectrum calculated with cloud layers between 10 and 0.1 bar pressures for the temperatures and radii indicated. The atmospheric windows corresponding to the J, H, Ks and L' bands are indicated in gray at the bottom of the figure. The solid circles indicate the magnitudes predicted by the synthetic spectra, the empty circles with error bars corresponding to the observations. From Marois, C. et al. (2008).

Phase curves and atmospheric circulation of exoplanets

When an exoplanet orbits around its star, it shows alternately its night side when it is close to the primary transit and its dayside near the secondary transit. By following very precisely the flux during the revolution of the planet (the phase curve), it is therefore possible to obtain information on the variations of its luminosity as a function of its position. The remark is also valid for exoplanets outside transit, provided the Earth is close enough to the plane of their orbit. The analysis was made using the Spitzer satellite for a number of hot Jupiters, and in particular HD 189733 b (Figure 8.11), which always presents the same face towards the star. The study showed that in all cases the hottest spot on the planet was not in front of the star, but was shifted eastward by some thirty degrees, indicating the presence of a violent equatorial wind, which could reach 1 km/s. This wind is predicted by some models of global climate circulation.

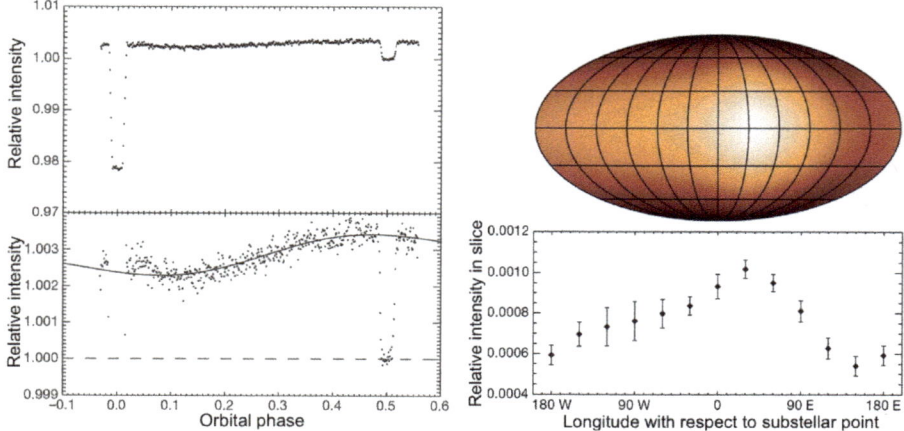

FIGURE 8.11 – Left: the phase curve of the exoplanet HD 189733 b (top), and an enlargement to show the variations of the emission by the planet (bottom). The orbital phase ranges from 0 to 1 during the period of revolution of the planet (here 2.22 days). On the right, a map of the brightness of the planet as a function of the longitude, derived from the phase curve. The offset from the point facing the star results from zonal winds at the equator. From Knutson et al. (2007).

The phase curves of the hot Jupiters also make it possible to measure the difference in temperature between the day and night sides. A study of half a dozen objects showed an increase in this difference with the average temperature of the exoplanet, in agreement with radiative transfer models

The physical nature of exoplanets   121

(Figure 8.12): the warmer the planet, the higher the contrast between the night side and the day side. But it is possible to go even further: by analyzing the phase curves of several exoplanets very close to their stars and using models of global circulation, one can deduce the thermal profiles of the day and night sides. For now, we only have to do that the photometric measurements made by the Spitzer satellite at 3.6 and 4.6 µm. By 2018, thanks to the JWST, and beyond, with dedicated observatories, it will be possible to obtain complete spectra along the phase curve and to identify the atmospheric constituents.

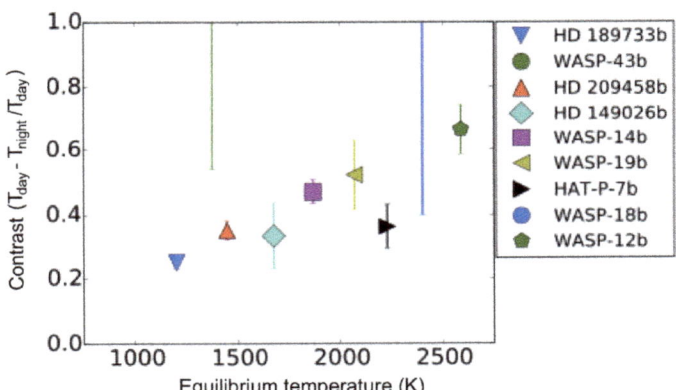

FIGURE 8.12 – Measurement of the relative temperature difference between day and night, as a function of the equilibrium temperature of the exoplanet. The vertical lines represent possible ranges for two exoplanets. We see that this difference increases with the equilibrium temperature. The selection consists of hot Jupiters in synchronous rotation, which always shows the same face towards the star. From Parmentier, V. et al., with thanks.

Bibliography

Tinetti, G. & Griffith, C.A. (2010) Exploring Extrasolar Worlds Today and Tomorrow, *ASP Conference Series* 430, 115. http://cdsads.u-strasbg.fr/abs/2010ASPC..430..115T.

Forget, F. & Leconte, J. (2014) Possible climates in terrestrial exoplanets, *Philosophical Transactions of the Royal Astronomical Society* 372, 20130084. http://cdsads.u-strasbg.fr/abs/2014RSPTA.37230084F.

Madhusudhan, N. et al. (2016) Exoplanetary Atmospheres – Chemistry, Formation Conditions, and Habitability, *Space Sciences Reviews* 205, 285.

Chapter 9
Around Exoplanets

Our Solar System contains, in addition to planets, many small objects that astronomers call "small bodies": asteroids, the icy objects of the Kuiper Belt, comets and the satellites and rings of the planets themselves. It is therefore legitimate to ask whether such bodies exist in exoplanetary systems. As for asteroids and icy objects, their direct detection is problematic since it is already difficult to discover planets similar to the Earth. However, it is likely that there are such objects in most planetary systems, which are part of the disk or debris belt that can often be seen in the infrared or in millimeter and sub-millimeter waves (Figure 9.1).

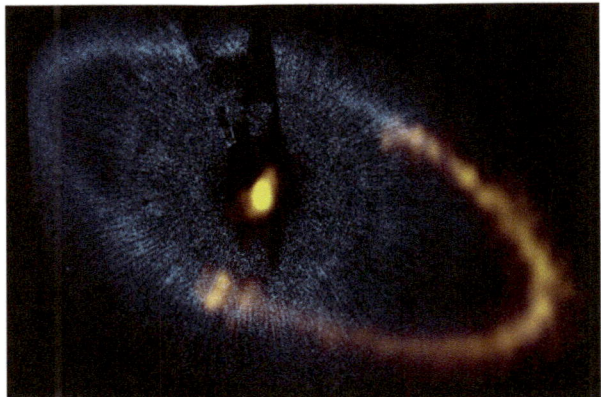

FIGURE 9.1 – The star Fomalhaut A is surrounded by a debris belt, revealed here in this composite image by the dust thermal emission at 0.86 mm wavelength observed with the ALMA interferometer (red) and by the light scattered by dust observed with the Hubble Space Telescope (blue). The dark areas and the blue area around the central star are artefacts of the optical observation. The belt is at 140 a.u. from the star, and its edges are relatively sharp, suggesting that this belt is in gravitational interaction with two planets, one of which is known (Fomalhaut A b). It is the equivalent of the Kuiper Belt of icy objects of our Solar System, but with dimensions 4 times larger. © ALMA (ESO/NAOJ/NRAO) and NASA/ESA.

Some large asteroids may appear as gravitational microlenses during specialized researches that has already revealed about 50 exoplanets (see Chapter 3). Large satellites similar to the Galilean satellites of Jupiter or Titan could also manifest themselves in this way, as a small peak of luminosity accompanying that produced by a planet (see Figure 3.8). But for the moment nothing of this has yet been observed. However, the transit of a satellite has perhaps been seen, as described later.

The situation is more favorable for comets, as we will see, and a giant ring has been discovered around an object that could be an exoplanet.

The exocomets

The first exocometers were discovered around the star of the southern hemisphere β Pictoris. The system of this star includes the massive planet β Pictoris b (Figure 9.2), and perhaps two other planets as well as an important disk of debris seen edge-on.

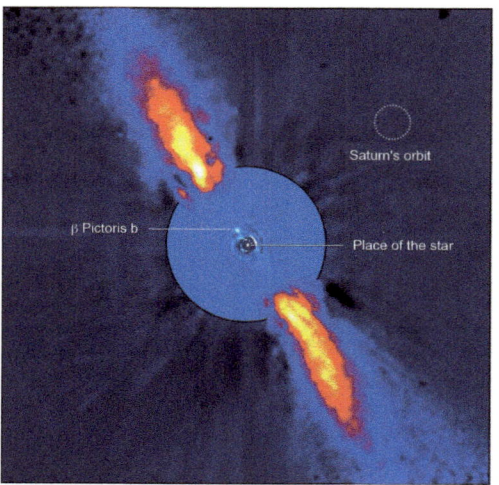

FIGURE 9.2 – The star β Pictoris and its exoplanet β Pictoris b, observed with adaptive optics with ESO's Very Large Telescope in October 2003. At the center, the position of the star itself, occulted by a coronograph. The orbit of Saturn, whose radius is 9.5 a.u., is indicated for comparison. The image of the debris disc, observed in the near infrared, is superimposed (false colors indicate the intensity). © ESO.

If the system of β Pictoris contains comets, they cannot be seen directly because they are embedded in the intense light of the star. But when a comet passes in front of the star and loses matter, the ejected gas can be detected by spectroscopy because it produces absorption lines in the spectrum of the star at well-defined wavelengths. Such lines have been seen for some twenty years by a group of researchers from the Institut d'Astrophysique of Paris (Figure 9.3).

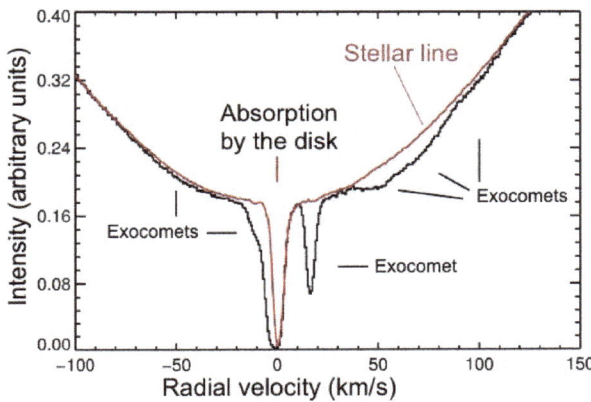

FIGURE 9.3 – Profile of the ionized calcium line at the wavelength of 393.4 nm observed in the direction of β Pictoris. There are several superimposed absorptions. The wavelength λ is transformed into radial velocity v (projection of the velocity on the line of sight) by using the Doppler-Fizeau effect: $v = c(\lambda - \lambda_0)/\lambda_0$, where c is the velocity of light and λ_0 a reference wavelength, corresponding here to the radial velocity of the star. The disk produces a time-independent narrow absorption on the very broad line of the star (in red). Other temporary absorptions (in black) are due to the ionized calcium present in the gas ejected by comets passing in front of the star, with various velocities. This spectrum was taken on 27 October 2009 with a high-resolution spectrograph placed at the focus of the 3.60 m diameter telescope at the ESO Observatory in La Silla, Chile. According to Kiefer et al. (2014).

This team has counted no less than 6000 passages of comets in front of β Pictoris, enough to make excellent statistics. On average, six different comets pass at the same time in front of the star at different distances. The observed spectral lines, which are the two lines of ionized calcium at 393.4 and 396.8 nm, provide very valuable information on each of these comets. The intensity of the lines makes it possible to calculate the fraction of the surface of the star which is covered by the coma (the gas envelope) of the comet, as well as the quantity of gas that it contains. From this we deduce the "efficiency" of the comet to eject gas, and thus also dust. The model also provides the distance from the comet to the star at the moment of its passage; it varies between 0.01 and 60 astronomical units according to the comet.

The result of the statistics is quite surprising. There are two different types of comets (Figure 9.4). Some of them, which are losing a lot of gas and dust, have a very extended coma that sometimes almost completely covers the star when they pass in front of it; they have orbits rather similar to one another. They are perhaps fragments of one or more rather massive comets: they are therefore "fresh", very active objects. The other comets produce much less gas and dust and have a smaller coma: they are therefore older objects, which have already passed many times near the star. These last comets undergo strong gravitational perturbations due to the planet β Pictoris b. They generally pass closer to the star than those of the first type, and have very elongated orbits due to these disturbances, which causes them to spend much time at large distances from the star. Would a reservoir of comets analogous to the Kuiper belt be formed in this way? We still have no answer to this question.

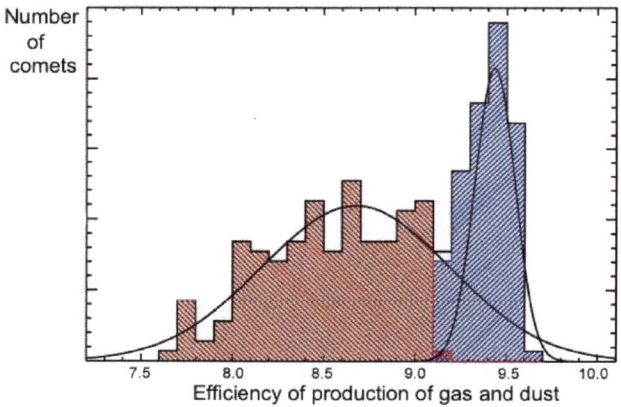

FIGURE 9.4 – Efficiency of gas and dust production by the comets surrounding the β Pictoris star. This quantity is proportional to the logarithm of the amount of matter which would be emitted by the different comets supposed to be placed at the same distance from the star; it is therefore independent of the distance of the comet from the star. Two populations are clearly visible. In blue, very active young comets, which may result from the rupture of one or more large younger comets. In red, older and less active comets, which have very eccentric orbits, and have already passed many times near the star. According to Kiefer et al. (2014).

The β Pictoris system is observed at an early stage of its evolution, where the disk of debris that surrounds the star is still extremely active. A dozen other young stars are known to be also surrounded by comets, in particular HD 172755 and η Corvi, but there must be many others. The Spitzer and Herschel space observatories have observed the presence of water vapor around some young stars, up to a hundred or so astronomical units from the star: this is probably the result of the evaporation of comets. The Odin and

Herschel satellites also detected an abnormally high emission of water vapor around stars far more evolved than the previous ones. The phenomenon could be explained in this case by the massive evaporation of icy objects of a Kuiper belt around the star. These observations give direct confirmations of the models of evolution of the protoplanetary disks and of the formation of the planets, which have been presented in Chapters 6 and 7.

The debris disk of β Pictoris contains other gases than ionized calcium and water vapor, in particular carbon in atomic form, which is very abundant, iron, and the carbon monoxide molecule whose emission in millimeter waves was mapped by ALMA (Figure 9.5). These atoms and molecules are not part of the gas of the initial disk, which has been entirely evaporated, but come from the sublimation of the cometary material.

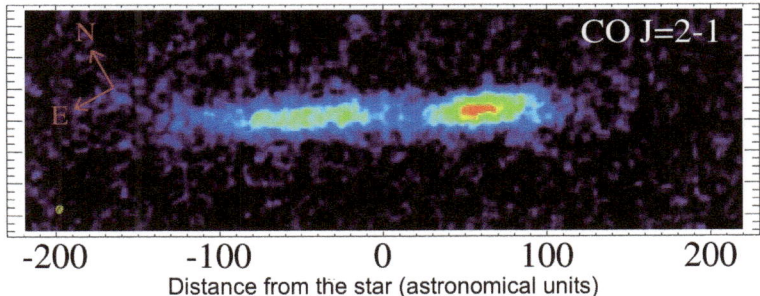

FIGURE 9.5 – The CO molecule in the debris disk of β Pictoris, observed with ALMA by its J = 2-1 rotation line at 1.3 mm wavelength. The angular resolution is represented by the small circle at the bottom left. From Matrà, L. et al. (2017) *Monthly Notices of the Royal Astronomical Society* 464, 1415.

Giant rings around an exoplanet?

All four giant planets of the Solar System are surrounded by a system of very flat rings, of which that of Saturn is known from Galileo and Huygens. There is no reason why this should not be the same for giant exoplanets, and one can hope to detect them by the additional absorption they might produce if their planet passes in front of its host star. These rings are probably what remains of circumplanetary disks, where the satellites have been formed in the same way as the planets in circumstellar disks. They should therefore be particularly spectacular in very young planetary systems.

While the rings of Saturn are only partially opaque, the younger circumplanetary disks must be more opaque, and they can be expected to strongly absorb the light of the host star if they pass in front of it. We know several cases where a star is occulted at regular intervals by a disk, but these are binary stars, and the disk in question is the circumstellar disk that surrounds one of the stellar components. There is however a different case: that of a very young star, about 16,000 years old, with the barbaric name of 2MASS J14074792-3945427, abbreviated as J1407. It presented a very deep, irregular, long-lasting eclipse (Figure 9.6). It can be interpreted by the passage in front of the star of a disk shown schematically in Figure 9.7. This disk cannot be a circumstellar disk around a rather massive companion, for one would expect in this case the disk to be sufficiently heated by this companion to emit in the infrared: no infrared excess is observed in this system. The companion that owns the disc must therefore be a giant planet or a brown dwarf. New observations, and in particular that of another transit of the disk in front of the star, are necessary to define the mass and nature of this companion. It would also be interesting to discover similar cases, in particular by examining in detail the transits of massive planets in front of their host star.

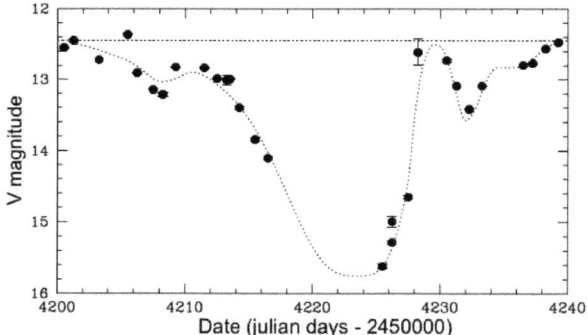

FIGURE 9.6 – Eclipse of J1407. The date in Julian days is indicated on the abscissa, and the ordinates give the magnitudes of the star, which are a logarithmic scale of luminosity such that 2.5 magnitudes correspond to a factor 10. We see that the eclipse is deep, long and complex. If it is indeed due to the circumplanetary disk schematically shown in Figure 9.7, it is not possible to observe the transit of the planet itself in front of the star because it is obscured by the opaque central part of the disk. From Mamajek *et al.* (2011).

FIGURE 9.7 – A possible model of the circumplanetary disk formed of concentric rings, responsible for the eclipse of the J1407 star. This star is represented on the right, at the right scale. The disk, whose dimensions are uncertain but of the order of several hundred times the diameter of Jupiter, has passed horizontally in front of the star. From Mamajek *et al.* (2011).

A satellite?

A systematic search was conducted recently for big satellites around transiting exoplanets. Multiple transits of 284 exoplanets discovered by the Kepler satellite, orbiting at 0.1 to 1 a.u. from their star, were examined in order to detect possible anomalies in the light curve that would be due to the transit of a satellite. As a result, there is a hint that satellites of the size of the Galilean satellites of Jupiter might exist very close to their planet, but conspicuous anomalies have been seen in only one case, that of Kepler-1625 b (Figure 9.8). In this case, a super-Jupiter orbits an old solar-mass star that is becoming a red giant. If the anomalies are not an artifact or due to some other cause, they can be explained by the transit of a satellite with a size similar to that of Uranus or Neptune, which orbits the planet at a distance of 19 times its radius, in about 9 hours. This short period comparable to the transit time of the planet in front of the star, about 20 hours, explains the complex appearance of the transit curves. However, the existence of the satellite and the conclusions above are only tentative, and further observations are necessary for a confirmation. The existence of such an enormous satellite was not expected. If confirmed, its discovery would raise new questions about the formation and survival of planetary satellites.

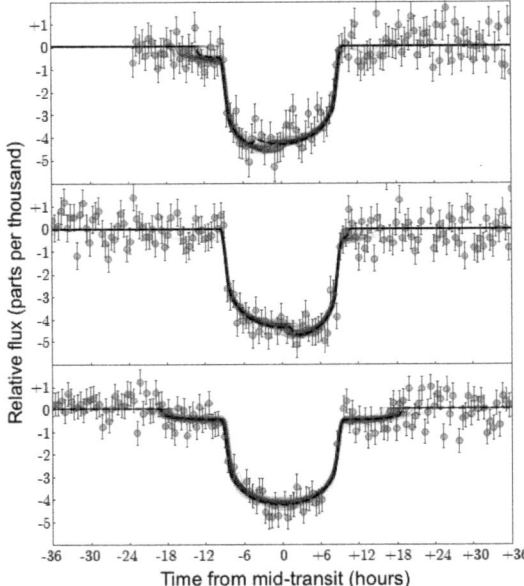

FIGURE 9.8 – The three transits of Kepler-1625 b observed by the Kepler satellite. The departures from a normal transit curve could be interpreted by the transit of a satellite of this exoplanet. The dark curves result from a number of simulations. From Teachey, A. et al. (2017).

Bibliography

Mamajek, E.E. et al. (2012). Discovery of an extrasolar ring system transiting a young Sun-like star, *Astronomical Journal* 143, 72.

Kiefer, F. et al. (2014). Two families of exocomets in the β Pictoris system, *Nature* 514, 462.

Teachey, A, Kipping, D.M. & Schmitt, A.R. (2017) HEK VI: on the Dearth of Galilean Analogs in *Kepler*, and the Exomoon Candidate Kepler-1625B I, *arXiv:1707.08563v1*.

Chapter 10
Life on exoplanets?

The burning question that everyone is asking about exoplanets is the possible presence of life on some of them. Unfortunately, we cannot answer this question yet, but we may be optimistic: we might have some answers in the coming decades. Meanwhile, a new discipline of astronomy is in full development: exobiology.

Exobiology, which is the study of living matter outside the Earth, or more modestly for the time being of its precursors, was born in 1920, when the biochemists Alexander Oparine (1894–1980) and John Haldane (1892–1964) expressed the concept of "primitive soup", from which complex molecules would have emerged under the effect of an energy supply. In 1952, the famous experiment of Stanley Miller (1930–2007) and Harold Urey (1893–1981) proved that it was possible to synthesize amino acids, the basic bricks of life, from simple molecules in the presence of water and under the effect of a source of energy such as electric sparks or ultraviolet radiation (see Box). Since then, other laboratory experiments have confirmed and amplified this discovery.

Astronomers have therefore begun to search for organic molecules, called *pre-biotic molecules,* which can be integrated into chemical reactions leading to the synthesis of "living" matter. The first medium they have considered is the interstellar medium itself: today, nearly 200 molecules in gaseous form or frozen on dust grains have been found there, mainly by spectroscopy at sub-millimetric to decimetric wavelengths. They comprise up to thirteen atoms, of which carbon is predominant. Particular macromolecules, the fullerenes C_{60} and C_{70} with a football-like geometry, and aromatic polycyclic hydrocarbons composed of tens or even hundreds of C and H atoms have also been found in abundance by spectroscopy in the infrared. All these molecules are found in dense molecular clouds, and some also in envelopes of evolved stars. Amino acids have also been searched for in the interstellar medium, but their detection is very difficult and no convincing results have yet been obtained.

Other sites within the Solar System also contain pre-biotic molecules: the atmosphere of Titan, Saturn's largest satellite, in which the Voyager space mission, supplemented by measurements taken from the Earth in millimeter waves, discovered in 1980 hydrocarbons and also nitriles such as HC_3N and CH_3CN, the first links of a more complex chemistry. Meteorites and comets harbor more or less complex organic molecules, including long carbon chains and amino acids; some were probably already present in the interstellar cloud from which the Solar System was formed, and others have been synthesized since. So we do not start from scratch, and we can even ask ourselves whether life on Earth has not originated in the fall of comets or meteorites that could have brought pre-biotic molecules: it is the very attenuated form of an old theory, *panspermia*, which claims that life on Earth would have come from elsewhere: we will discuss this later. That being said, no living beings have been found in those sites, and their possible arrival from other planets belongs entirely to the domain of myth.

We will now delve deeper into the problem of the origin of life and its possible presence outside the Earth in the Solar System and in exoplanetary systems. But we must first try to define what life is.

What is life?

It may seem strange to ask this question while we are surrounded by living beings, and we are such beings ourselves. But life has begun somewhere in a primitive form, and the question is rather: how and in what form did it appear? We know only one example of life: the life on Earth, and all our concepts are necessarily influenced by this example. But terrestrial life may be only a special case, just as the Solar System appears to be a rather peculiar case among the planetary systems; would it be only one element of a larger ensemble that we do not suspect? In this domain, as in the case of exoplanets, we must guard against a vision that would be too anthropocentric.

That said, scientists, who are not creationists, believe that life results from the action of the laws of physics and chemistry on appropriate materials, if the environment is favorable. Life would thus be the natural result of the evolution of certain chemical systems placed under favorable conditions, which can be extremely varied as evidenced by the presence of an autonomous life in deep ocean thermal springs and the presence of bacteria in quite unexpected environments. We do not possess an experimental verification of this idea, but the experiments we have cited are a step in this direction. This hypothesis is the basis of all research in exobiology, although it is not always explicit.

The definition of life is most often based on the main functions of the living beings: reproductive capacity, ability to use environmental energy, differentiation of the environment, and reduction of entropy by the creation of an organization. However, this definition is insufficient. By their formation, their growth and their multiplication in a solution, the crystals indeed

Life on exoplanets? 133

meet all these criteria, and nobody will call them alive. We could also use a biochemical definition, such as the presence of DNA. But is it certain that living beings on an exoplanet use DNA for their genetic code, or else another type of molecule? In any case, the living organisms must contain genetic material.

Exobiologists ultimately tend to define life by the possibility of Darwinian evolution. This implies, in addition to the above criteria, the existence of mutations and natural selection. Organisms must also be mortal, otherwise evolution is impossible. The unicellular organs that are certain algae meet all the criteria we have mentioned, and represent on the Earth the simplest forms of living matter. These forms are already complex and elaborated with their walls, proteins and possibly a nucleus.

The emergence of life on Earth

We might, of course, imagine that the first living organisms on the Earth were brought by comets or meteorites, and that others have thus arrived at different epochs: this is the theory of panspermia, which had its hour of glory. It is based on the fact that some bacteria appear resistant enough to survive a long journey in space. But it only displaces the problem, for it was then necessary for life to appear elsewhere, for example on Mars, from which we occasionally receive a fragment expelled by the fall of an asteroid or a large meteorite.

Let us forget panspermia and see how life could have emerged on the Earth. When the Earth was formed, the physical conditions were infernal, and water vapor and the other volatile molecules had been expelled from the protoplanetary disk by the evaporation due to the heat of the proto-Sun. The first solid elements, zircon crystals found in Australia, appeared on the Earth's surface after about 150 million years, followed by other rocks. Various arguments suggest the presence of water on Earth at that time. Where did it come from? It is believed that some of the planetoids that made up the Earth contained water and other volatile elements because they had been formed at a large distance from the proto-Sun and had migrated into the formation zone of the rocky planets (see Figure 7.2). Water could also have been brought after the formation of the Earth by a bombardment of comets and meteorites. Oceans were therefore present fairly early in the evolution of our globe. The Great Late Bombardment, which occurred about 800 million years after the Earth's formation, has disrupted the Earth's surface, partially evaporated the oceans and may have destroyed any form of life that might have appeared before. But it seems that the Earth's water and atmosphere have partly survived this episode.

What was this atmosphere made of? It has long been considered that it was formed from the gas of the primitive interstellar matter, which was strongly reducing because it consisted mainly of hydrogen. But we saw in

Chapter 7 that this gas was driven out very early during the formation of the Solar System, leaving only the refractory solid elements that were to constitute our globe. This is confirmed by the analysis of the isotopes of rare gases in the current Earth's atmosphere, whose ratios are different from those in the Sun, which reflect the ratios in the primitive interstellar gas. The atmosphere has thus formed from the water, carbon dioxide CO_2 and nitrogen included in some planetoids or possibly brought by the fall of meteorites and, to a lesser extent, comets. A considerable greenhouse effect, mainly due to CO_2, initially brought the temperature of the Earth's surface to a value of around 200 °C although the young Sun was 30% less luminous than today. As for the oceans, their temperature has decreased in the last 3.5 billion years from about 65 °C to the current temperature (Figure 10.1).

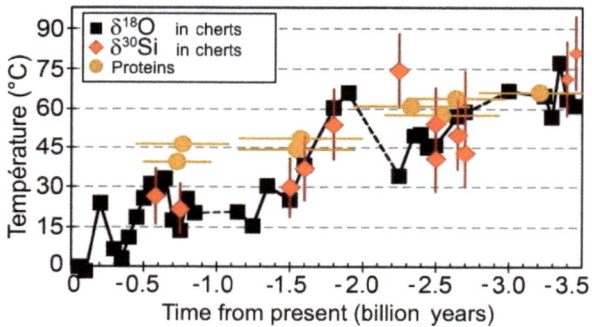

FIGURE 10.1 – Decrease of the ocean temperatures during the last 3.5 billion years. The present time is to the left of the abscissa scale. The temperature is evaluated from the $^{18}O/^{16}O$ and $^{30}Si/^{28}Si$ isotopic ratios in cherts, which are old siliceous sediments. Variations in these ratios are expressed as $\delta^{18}O = [(^{18}O/^{16}O)/(^{18}O/^{16}O)_{present} - 1] \times 1000$, the same for silicium. The $^{18}O/^{16}O$ isotopic ratio, measured in deposits of organic matter in rocks derived from protein degradation, is also indicated, confirming the other results. According to Tartèse, R. et al. (2017).

The decrease in the Earth's temperature was due to the weakening of the greenhouse effect, which was itself due to the reduction of the amount of CO_2 in the atmosphere, the convective motions of the Earth's mantle having stored a lot of CO_2 as carbonates. 2.3 billion years ago, the Earth was even completely frozen, and traces of this glacial phenomenon can be found in different parts of the world. This episode did not translate into ocean temperatures, either because it lasted a short time, or because it only interested the land.

But the first bacteria had already been born for a billion years. They were methanogenic bacteria, which derived their energy from the synthesis of methane, which had contributed to the greenhouse effect. Hundreds of million years later appeared the cyanobacteria (unicellular blue algae), which produced oxygen by photosynthesis; this oxygen destroyed the methane, which reduced the greenhouse effect and contributed to the lowering of the temperature of the Earth's surface. Note that the Earth's atmosphere and temperature were already affected by the effects of life, which is still true today.

The first bacteria, the traces of which are found in some dated rocks, appeared just over a billion years after the Earth was formed in, an atmosphere dominated by carbon dioxide, and in the presence of water. The necessary organic molecules may have been synthesized as in the Miller-Urey experiment, although the latter started from a very different reactive environment than the atmosphere of the time. However, the bacteria might be born in a locally reducing medium. Other possibilities exist. An example is provided by the hydrothermal sources in the Atlantic and Pacific ridges, where chemical reactions produce pre-biotic molecules from hydrogen sulphide SH_2, dissolved CO_2 and mineral nitrogenous substances: they contain colonies of bacteria. The volcanic hot springs on the Earth's surface are other possible sites for the origin of life. Despite these very different conditions, the pre-biotic molecules are always based on carbon. It has sometimes been considered that such molecules could be based on silicon, which is tetravalent like carbon, but silicon is much less abundant than carbon, and especially its chemistry is much poorer: among the some 200 molecules discovered in the interstellar medium, only a few contain silicon, while most of them contain carbon. This idea must be abandoned.

A very serious possibility is the supply of pre-biotic molecules by asteroids and comets. A recent study based on the chemical composition of the 67P/Churyumov-Gerasimenko comet observed by the European Rosetta probe suggests that, while the comets did not bring all the water on the Earth, they were able to provide a considerable amount of pre-biotic material, especially during the Great Late Bombardment. The mass of this material could be equivalent to all the biomass present today on our globe (about $2 \; 10^{12}$ tons). What is it made of? In comet 67P no less than 58 different molecules were detected, mainly organic, including an amino acid, glycine, and some products particularly chemically active, such as hydrocyanic acid HCN and formaldehyde H_2CO. The comet also contains a large amount of refractory organic material, perhaps mainly polyoxymethylene, a macromolecule that results from the polymerization of formaldehyde, and a similar amount of mineral matter, mainly silicates. Some meteorites, the carbonaceous chondrites, contain material apparently similar to cometary matter but annealed, and could have participated in seeding the Earth.

Once the amino acids had been synthesized or arrived on the Earth, proteins had to be made from them, starting with the peptides that are formed by the combination of two of these acids (Figure 10.2). Then, it is easy to group the peptides into polypeptide chains. The proteins are polymers of such peptide sequences, where different amino acids are arranged in a defined order. Only 20 amino acids are involved in terrestrial living matter, whether animal or vegetal. For example, the protein of the egg white corresponds to a polypeptide chain of 129 molecules belonging to these 20 amino acids. Some of these chains can be spirally wound, a structure favored by the presence of hydrogen bonds between the NH group of one peptide unit and the CO group of another.

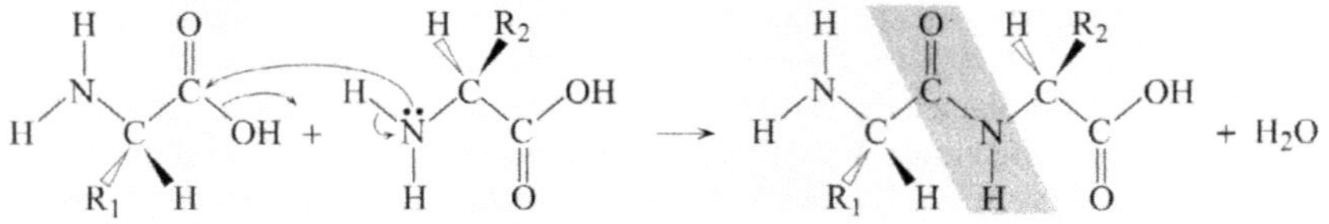

FIGURE 10.2 – Formation of a peptide bond by combination of two amino acids. R1 and R2 are characteristic radicals of each amino acid (for example H for glycine, CH_2OH for serine). The arrows to the left indicate the detailed mechanism of reaction. The amide group -CO-NH- in gray is the peptide bond which connects the two sections. From R. Luft (2014).

Note that while nearly 80 different amino acids have been found in meteorites and comets, they include only 8 of the 20 that are needed for terrestrial life. We do not know where the others come from. Note also that no peptides have been found in the interstellar medium, comets and meteorites, which implies that the synthesis of the peptides takes place under particular circumstances: different possibilities have been imagined to provide the necessary energy, including the fall of a comet on the Earth's surface, or more simply a reaction on a clay support, which was actually demonstrated in the laboratory as early as 1970.

While proteins play a large role in living cells as hormones, enzymes, tissue supports (collagens), etc., they are not the basic elements for the formation of nucleic acids, which contain the genetic information. The nucleic acids are built from four so-called nucleotide molecules, containing phosphorus and nitrogen (Figure 10.3). Atomic phosphorus, nitrogen and the phosphorus and nitrogen molecules that are required for nucleotide synthesis are found in comets and the interstellar medium. The nucleotides have structural analogies with the peptides, and are like them capable of assembling themselves into long helical chains: the ribonucleic acid RNA and the deoxyribonucleic acid DNA (Figure 10.4), which is probably formed from RNA. It is known that DNA, which contains genetic information, can reproduce itself by duplication. But we know nothing about of the initial formation of this complex macromolecule. Laboratory experiments have allowed us to reproduce only the very first stages, the formation of nucleotides: this is the whole problem of origin of life.

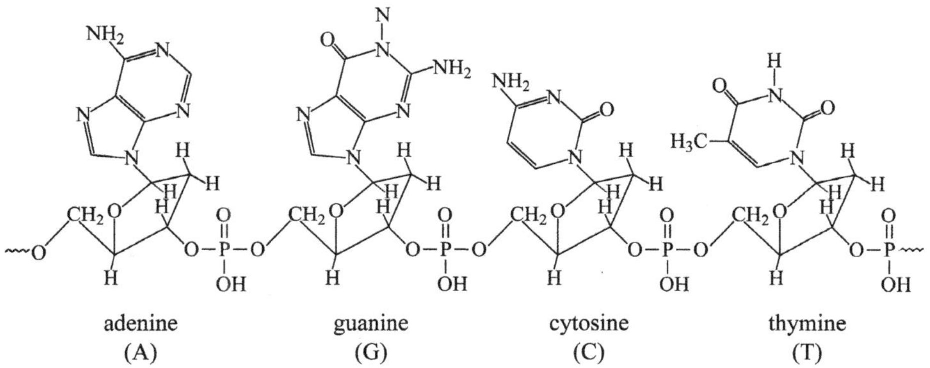

Life on exoplanets?   137

adenine guanine cytosine thymine
  (A)   (G)   (C)   (T)

FIGURE 10.3 – A basic structural fragment of DNA, containing the four linked nucleotides. Most carbon atoms are not represented to lighten the presentation. From Luft (2014).

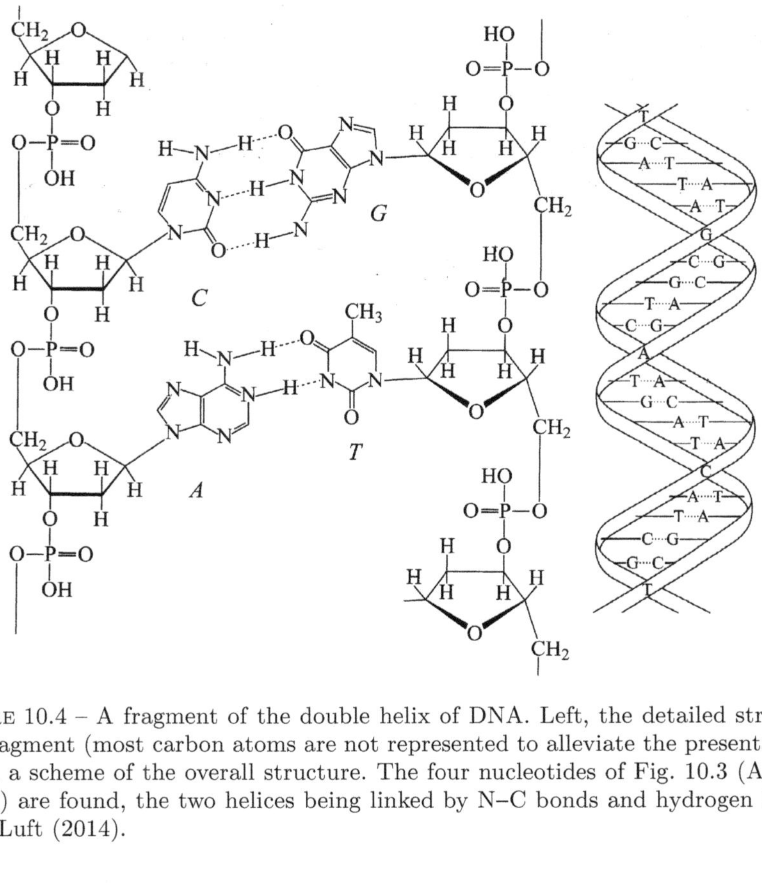

FIGURE 10.4 – A fragment of the double helix of DNA. Left, the detailed structure of a fragment (most carbon atoms are not represented to alleviate the presentation). Right, a scheme of the overall structure. The four nucleotides of Fig. 10.3 (A, G, C and T) are found, the two helices being linked by N–C bonds and hydrogen bonds. From Luft (2014).

In addition to proteins and DNA, the living cells such as mono-cellular algae have an envelope that separates them from the external environment and through which energy exchanges of products, and in particular nutrients, take place. This envelope is made up of lipids, which in living matter are very frequently phosphoglycerides (Figure 10.5). The lipids are amphiphilic: the hydrophilic end (the head) of the macromolecule, which has a dipole moment, grips the water molecules that also have a dipole moment. The other end (the tail) is hydrophobic. In the presence of liquid water, the lipids stick to each other perpendicular to the surface in such a way that the hydrophilic heads are in contact with water, the hydrophobic tails rising on the other side. A monomolecular film is thus formed on the surface of water.

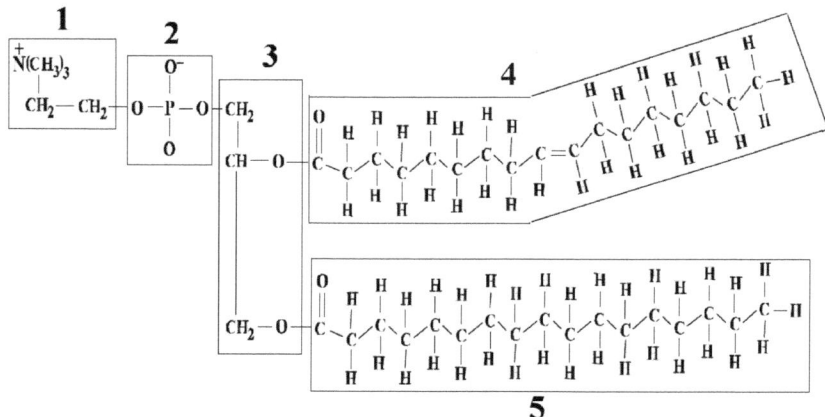

FIGURE 10.5 – Formula of a typical phosphoglyceride. Group 1 is choline, group 2 is phosphate and group 3 is glycerol. Together, they constitute the hydrophilic head, which possesses a dipole moment due to the + and − charges. The tails 4 and 5 are fatty acids, respectively unsaturated and saturated, which are hydrophobic. Wikimedia Commons, Danntzikg.

The lipid monolayers can spontaneously assemble into double flat or curved structures, as in Figure 10.6, thus separating an inner and an outer environment, as in a living cell. They are able to filter certain molecules and to retain others. Laboratory experiments made from the amino acids extracted from the Murchison meteorite, which contains nearly 80 of them, have seen more or less spherical vesicles similar to liposomes develop rapidly in water. We are not far from the cell walls!

Life on exoplanets?

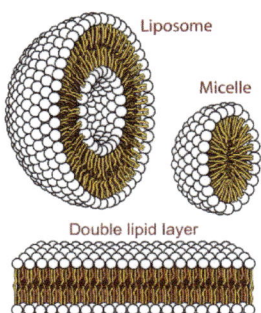

FIGURE 10.6 – Structure of lipid layers. The hydrophilic heads of the molecules are represented by white balls, their hydrophobic tails by yellow filaments. The monolayers can close on themselves forming micelles, which are frequent in colloidal solutions. The double layers can also close on themselves, forming liposomes which can be manufactured artificially. Wikimedia Commons, LadyofHats.

We must also understand not only how the actual cell walls can be formed, but also how they can trap macromolecules such as proteins or DNA. Experiments have shown that if a mixture of vesicles and macromolecules are subjected to hydration and dehydration cycles, imprisonment of these macromolecules may occur, since bonds are formed in replacement of those left free by the departure of water. This may occur in periodically desiccated pools, in particular those fed by volcanic hot springs. Finally, we must understand the origin of the metabolism which ensures the survival of the cells. We are still a long way off in this matter.

Another observation that poses problems is the chirality of pre-biological or biological molecules. What is it about? Just as our left hand cannot be superimposed on our right hand, but is its image seen in a mirror, some pairs of molecules have a similar property, chirality. For example, a molecule made of a carbon atom surrounded by four different atoms or substituents is chiral (Figure 10.7). Such a molecule has two isomeric forms which are the mirror image of each other. One of these isomers has the property of rotating clockwise the plane of polarization of light which passes through it and is said to be dextrorotatory, while the other rotates it in the other direction and is called levorotatory. However, the basic molecules of the terrestrial organic matter are always levorotatory when they are chiral, which is almost always the case for amino acids with the notable exception of glycine. In the case of the chiral pre-biotic molecules synthesized in the laboratory by experiments like that of Miller-Urey, it turns out that they contain the dextrorotatory and levorotatory isomers in equal amounts. This is not the case with the amino acids of meteorites: if they are roughly the same as in this well-known experiment, they are mostly levorotatory like terrestrial pre-biotic molecules. This is a strong argument in favor of an extraterrestrial origin of pre-biotic molecules, even if we do not understand well what initially favored the levorotatory form. However, recent laboratory experiments seem to indicate, contrary to older ones, that the synthesis of amino acids irradiated by a polarized ultraviolet flux (produced, for example, in a region of massive star formation) leads to an excess of levorotatory chirality in a proportion comparable to that measured in meteorites. This is an interesting track, but there are other possibilities.

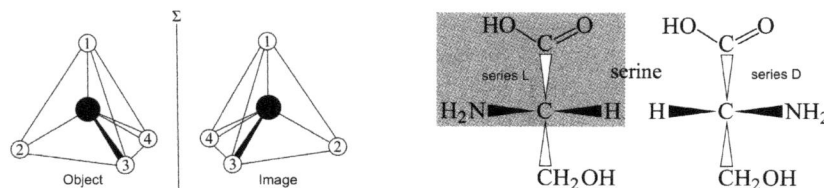

FIGURE 10.7 – Chiral molecules. On the left, the two molecules formed of a carbon atom with its four chemical bonds (the forward one is in black), surrounded by atoms or substitutes 1, 2, 3 and 4 different from each other. They are the mirror image of each other with respect to the plane Σ. On the right, the two isomeric forms of a chiral amino acid, serine. The forward links are in black. The left-hand form is levorotatory (series L) and the right-hand form dextrorotatory (series D). The gray rectangle indicates the reference pattern for chirality of amino acids, peptides and proteins. From R. Luft (2014).

To summarize, the emergence of life on Earth required a number of conditions: presence of water, carbon compounds, locally reducing medium, moderate temperature, and especially a long duration.

Once the first bacteria were born, evolution has built up more and more complex living beings from them (Figure 10.8). This took time: 2.3 billion years passed between the appearance of the first bacteria and that of the first multi-cellular beings, and another 650 million years before the explosion of life at the Cambrian period. In particular, it was necessary to synthesize chlorophyll, which carries out photosynthesis by absorbing CO_2 and produces the dioxygen which is presently an important part of the atmosphere: this happened about 500 million years after the first living cells. On the Earth, the evolution has been marked by climatic variations, continental drift, five episodes of extinction of many species, the origin of which is still discussed (perhaps volcanic), and most recently by the action of man. Finally, the increase in the temperature of the Sun will cause the Earth to become hardly habitable in 1.5 billion years, and even less if man persists in participating in the warming of our globe. There is no reason for the evolution after the creation of the first multi-cellular beings to have a similar history on other planets that would eventually harbor life, and we will not discuss it further.

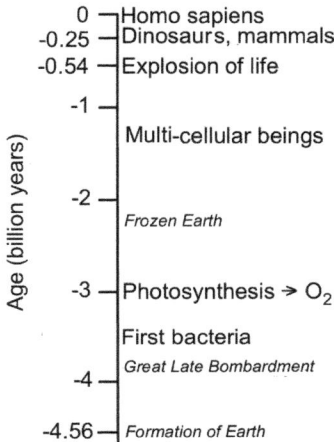

FIGURE 10.8 – Chronology of life on Earth. Diagram by the authors.

Life elsewhere in the Solar System?

Astronomers and the general public have long been interested in the possibility of life outside the Earth. The history of these investigations, where the marvelous has often dominated scientific reasoning, is too long and too complex to be told here. Finally, only two types of possible environments in the Solar System can be retained today: terrestrial planets (starting with Mars) and some satellites of the giant planets. Among these, two targets are currently privileged: Europa which is one of the four Galilean satellites of Jupiter, and Enceladus, one of the satellites of Saturn.

First, Mars. There is very little water presently in the atmosphere of Mars, but observation of its surface reveals the existence of rivers and other traces of aqueous flows in the distant past (Figure 10.9). The climatic conditions were therefore more lenient in the past, and became harder due to a major modification of its atmosphere: the escape of most of the atmospheric gases, which seems to be due to the erosion of the atmosphere by the solar wind. Water was only abundant a very long time ago, between -4 and -3.5 billion years. The presence of life born at this time, before the appearance of life on Earth, is possible. But the climatic conditions mean that life probably could not evolve much during the 500 million years when there was liquid water, so that it is expected to find on Mars, at best, only remains of uni-cellular beings. It could be that life has been brought to Earth from Mars: at this time when the bombardments on the planets were intense, the impacts often ejected fragments of Martian surface, some of which fell on the Earth. This still happens at present, but much more rarely. One of the meteorites found on Earth from Mars has been formed 4 billion years ago in a reducing and mild environment. It was believed to contain microorganisms, but one had to disillusion: it certainly contains organic matter, but at least part of it can come from contamination during its long stay in the Antarctic ice.

FIGURE 10.9 – Traces of ancient rivers on Mars, observed by the European Mars Express probe. © ESA/DLR/FU Berlin.

The discovery of traces of a past life on Mars would obviously be an important event on the philosophical as well as on the scientific level. To prove the absence of past life would also be interesting, because one would then wonder why, on two planets that have experienced similar conditions, one would have carried life and not the other. Finally, Mars offers much better archives than the Earth, where almost everything older than 3 billion years has been erased due to the drift of continental plates and erosion. For these reasons, Mars has been, and still is, the object of numerous explorations by space probes and rovers that move on its surface. No trace of life has been discovered until now. But one will search for fossil bacteria, organic matter residues or chemical anomalies due to life as a deficiency in the $^{13}C/^{12}C$ isotopic ratio. It will undoubtedly be necessary to dig rather deep in the Martian soil to find them.

Among the terrestrial planets, one can also ask the question of Venus. Today, the climatic conditions of the planet Venus are particularly inhospitable: the surface pressure is close to one hundred times the terrestrial atmospheric pressure, and the temperature, due to a runaway of the greenhouse effect, is above 450 °C. Moreover, it is covered with a thick layer of clouds made of sulfuric acid! However, the situation might have been radically different at the beginning of the planet's history. The solar flux was then less intense than today, about 70% of the present value. Assuming that the planet was the same distance from the Sun as it is today, its equilibrium temperature was initially about 300 K, or 30 °C. In other words, Venus was able to shelter a warm ocean at the beginning of its history! But, if it ever existed, this ocean had since evaporated (or boiled?), as the solar flux increased with time. And if ever life has appeared in the distant past, we shall never see its traces, for these have been erased by volcanism, whose age does not exceed a few hundred thousand years.

The case of Europa and Enceladus, which are respectively satellites of Jupiter and Saturn, is different from that of Mars and Venus. For Europa, the images sent by the Voyager probes revealed an icy surface, with a complex network of fractures (Figure 10.10). It could float over a viscous or even liquid ocean. The deformations by the tides due to Jupiter and its other satellites, tides whose amplitude reaches about thirty meters, produce a heating which would melt in the depths the ice covering the satellite. The ocean of Europa is probably salty: indeed, Europa has a magnetic field, perhaps induced by that of Jupiter, which implies the presence of an ionized and conductive medium. The water in this ocean could be in direct contact with the silicate nucleus of the satellite, making it possible to develop a complex chemistry that could lead to life even under apparently unfavorable temperature conditions that recall those of under-glacial lakes of Antarctica, where however colonies of bacteria thrive.

FIGURE 10.10 – The surface of Europa, seen by the Voyager 2 probe. The structure of fractured ice suggests the presence of solid plates floating on a liquid or viscous ocean. Some small and clear circular structures visible at several locations could be areas of collapse over lakes close to the surface. © NASA.

Enceladus also has an icy and fractured surface, and jets of material emerge from cracks where the temperature is higher than that of neighboring regions (Figure 10.11). These jets are geysers of fine particles of water ice and water vapor, coming from an underlying ocean heated by the tides due to Dione, another satellite of Saturn. More recently, hydrogen has been found in these jets. Here is another medium which might be appropriate to the appearance of life.

Let us note finally that it is not excluded that other satellites of the giant planets have niches favorable to the emergence of a primitive life. In the case of the Galilean satellites, Ganymede and Callisto could also shelter an ocean of liquid water beneath their surface. However, since they are farther from Jupiter than Europe, the tidal effects induced by the planet are smaller and, according to theoretical models, this ocean could be confined between two layers of ice. This would reduce the possibilities for a prebiotic life. The same is true for Saturn's Titan satellite. But other satellites might bring some surprises...

FIGURE 10.11 – The jets of Enceladus, seen by the Cassini probe. On the surface are some cracked fissures known as "tiger scratches", through which jets sometimes emerge. © NASA.

How to detect life on exoplanets?

Beyond the Solar System, it would be crucial to detect life on other planets. It is of course possible to wait for a possible signal, which would necessarily be emitted by an advanced civilization. This is the aim of SETI (Search for Extra-Terrestrial Intelligence) searches, which we will discuss in Chapter 12. But it would already be very interesting to detect signs of more primitive life. The terrestrial atmospheric oxygen is known to be the result of chlorophyll photosynthesis by plants, including cyanobacteria and uni-cellular algae, which are the most primitive forms of life. If these plants were to disappear suddenly, the oxygen would also disappear rapidly through combustion, or by oxidizing various minerals, and it is therefore necessary that it should be renewed. If oxygen is found on an exoplanet, there is a good chance that this planet harbors plants, at least uni-cellular ones. And there is good reason to believe that, as on Earth, oxygen is essential for the metabolism of more advanced beings that have high energy requirements, because only oxidation-reduction reactions involving oxygen can provide enough energy. One might also consider for such reactions other oxidizing gases such as fluorine, chlorine or bromine, but they have devastating effects when in contact with organic matter and are thus excluded. If there are living beings on a planet, its atmosphere must also contain CO_2 and methane; but the presence of these elements, which are very widespread, is by no means a proof of life. Although methane in the Earth's atmosphere appears to be largely biological, the detection of methane in an exoplanet would prove nothing in the absence of other indicators. It is oxygen that must be searched for.

Oxygen is in the form of dioxygen O_2, which is a symmetrical molecule whose spectral lines and bands are very weak. The only line that is possibly usable for detection in an exoplanet is in the red, at a wavelength of 760 nm. It is preferable to try to detect ozone O_3, which is produced by the effect of the ultraviolet radiation of the star on the dioxygen: although much less abundant, its spectral bands are stronger. In particular, there is a strong band of ozone at a wavelength of about 9.6 microns, which is well suited for detection. It is, of course, present in the spectrum of the Earth's atmosphere (Figures 10.12 and 10.13), so that the detection in exoplanets must take place, as in the oxygen band, from space telescopes.

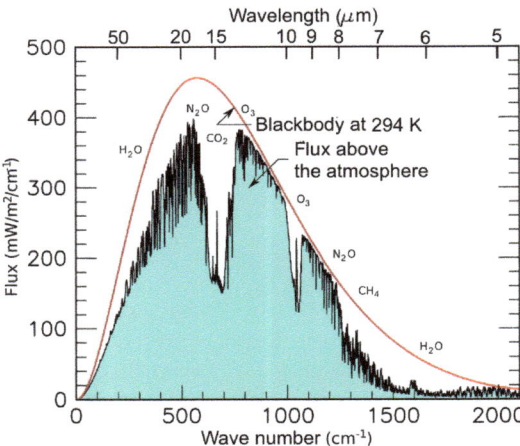

FIGURE 10.12 – The spectrum of the radiation emitted by the Earth, calculated from a model of Earth's atmosphere. In red, the emission in the absence of atmosphere, which would be practically that of a black body at 294 K. In blue, the actual emission, where we see the absorption by different atmospheric molecules, in particular of ozone at 9.5 μm. © NASA.

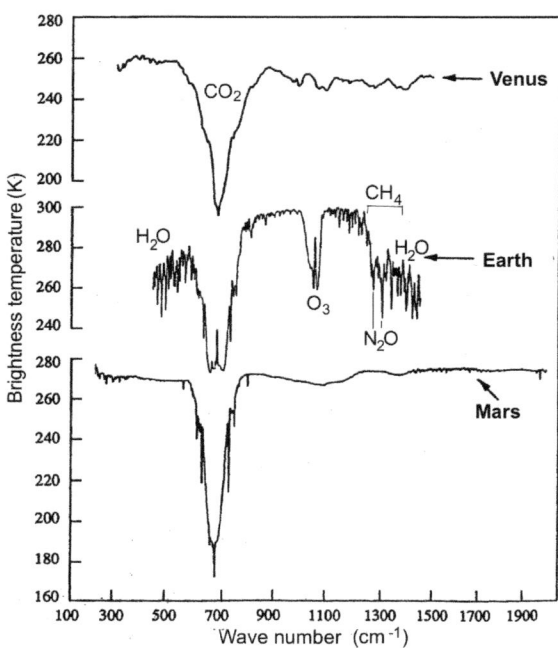

FIGURE 10.13 – Spectra of Venus, Earth and Mars in the mid infrared, observed by the Galileo probe en route to Jupiter between 5.2 μm (1900 cm^{-1}) and 33 μm (300 cm^{-1}). The flux is expressed as the brightness temperature, which would be independent of the wavelength if the source were a blackbody. In the case of Earth and Mars, the continuum corresponds to the temperature of the surface. In the case of Venus, it is the temperature of the top of the cloud layer of sulfuric acid, at 60 km altitude. The spectra show the characteristic signatures of the atmospheric gases: for Mars and Venus, CO_2 and a very small amount of H_2O and for the Earth CO_2, a lot of H_2O, and in addition ozone O_3, nitrogen oxide N_2O and methane CH_4 which are of biological origin: N_2O is mainly produced by bacteria and nitrophilic algae, and methane comes from bacteria, ruminants and plants. Adapted from Hanel, R. et al. (1992) *Exploration of the solar system by infrared remote sensing*, Cambridge University Press.

Chlorophyll shows two main absorption bands in visible light at 450 and 650 nm (Figure 10.14). The second band produces a discontinuity of a few percent (depending on the vegetation of the studied area) at 700 nm in the scattered light of the Sun. Its observation, already difficult for the Earth, would be almost impossible in an exoplanet. Moreover, the assimilation of CO_2 by living beings could be done via other molecules whose spectrum would be different.

Whatever method is used to detect life, life can only exist on a planet in the habitable zone. Such planets are necessarily not far from their star, and it would be very difficult to see them directly and a fortiori to obtain their spectrum (however, some young planets far from the star could be temporarily habitable at a certain stage of their cooling). The only practical possibility is to observe from space the transits of habitable exoplanets, which could be Earths as well as super-Earths (see Chapter 8). In order to do this, we must wait for the launch of the JWST, the successor to the Hubble Space Telescope, which is foreseen in 2018: it should be able to detect the O_3 band at 9.6 µm, if present, with its infrared spectrometers (Figure 10.15).

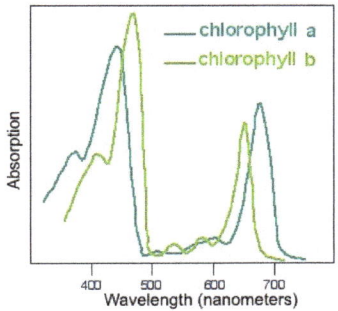

FIGURE 10.14 – Absorption spectrum of the two chlorophyll species obtained in the laboratory. From http://www.snv.jussieu.fr/bmedia/Photosynthese-cours/04-pigments.htm.

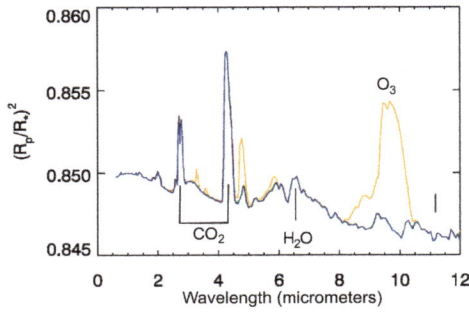

FIGURE 10.15 – Simulated spectrum of the terrestrial exoplanet TRAPPIST-1 d, which is perhaps in the habitable zone, during primary transits in front of its star of type M8. It is presented in the form of the variation with the wavelength of the apparent surface of the planet (R_P and R_* are respectively the radius of the star and that of the planet): an absorption in the atmosphere produces an increase in the apparent size of the planet. The simulation assumes an atmosphere similar to that of the Earth, without ozone in black, and with the terrestrial amount of ozone in yellow. The vertical bar on the right is the error bar ($\pm 1\sigma$) of an observation at 9 µm with the MIRI spectrograph of the JWST, after the accumulation of 90 transits. Ozone would be detectable if its abundance is greater than about 1/100 of terrestrial abundance (note that the effect of its abundance on R_P is not linear at all). From Barstow J.K. & Irwin, P.G.J. (2016) *Monthly Notices of the Royal Astronomical Society* 461, L92.

Box: The Experiments of Miller and Urey

In 1953, the famous experiment of Stanley Miller and Harold Urey proved that it was possible to synthesize amino acids from simple molecules, in the presence of water and under the effect of an electric discharge (Figure 10.E1). After a few days they found a change in the color of the condensed liquid, and after two weeks of continuous operation 10 to 20% of the initial carbon had formed organic products, including five amino acids, but neither sugars nor lipids or nucleic acids. Later, the sparks were replaced by ultraviolet radiation, with similar results. In 2007, after Miller's death, the contents of sealed containers containing the products of these experiments were reanalyzed by one of his students, Geoffrey Bada, who found that at least 25 different amino acids had been synthesized.

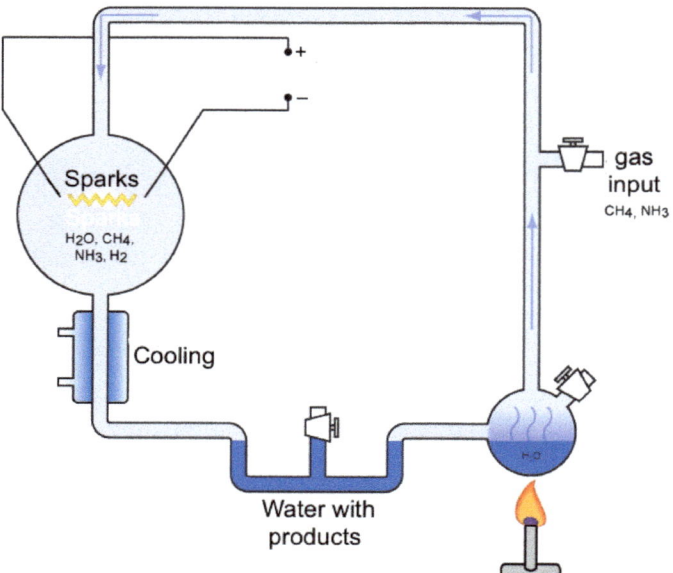

FIGURE 10.E1 – Diagram of the experiment of Miller and Urey. A mixture of water vapor, methane and ammonia is submitted to electric sparks. Hydrogen is produced, and various organic molecules are synthesized, which are collected in the condensed water. Wikimedia Commons, Carny.

Miller realized several variants of his experience, including one where a jet of steam was sent into the electric discharge: this was supposed to simulate the conditions of a volcanic eruption. It produced 22 amino acids, 5 amines and many molecules with hydroxyl radicals, more than the other experiments. In another experiment in 1961 by Juan Oró (1923–2004), hydrogen cyanide HCN, a very abundant molecule in the Universe, in aqueous solution with ammonia, produced amino acids and adenine, one of the four nucleotides of DNA; the other three nucleotides were synthesized later in similar experiments.

> All these experiments have used a reducing medium, and it has long been thought that the primitive atmosphere of the Earth was not reducing at the time of the appearance of life. This statement, however, is somewhat controversial, and it may be that some volcanic sites, for example, have been chemically reducing. In any case, the experiences of Miller and Urey represented a major step in the study of the origin of life.

Bibliography

Sagan, C. et al. (1993) A search for life on Earth from the Galileo spacecraft, *Nature* 365, 715.

Forget, F., Costard, F. & Lognonné, P. (2006) *La planète Mars, histoire d'un autre monde*, 2ᵉ ed., Paris, Belin.

Hamdani, S. et al. (2006) Biomarkers in disk-averaged near-UV to near-IR Earth spectra using Earthshine observations, *Astronomy & Astrophysics* 460, 617.

Maeder, A. (2012) *L'unique Terre habitée? Les conditions pour la vie sur les planètes*, Lausanne, Favre.

Coustenis, A. & Encrenaz, T. (2013) *Life beyond Earth*, Cambridge University Press.

Luft, R. (2014) *Biosphère et chimie, un laboratoire naturel*, collection Grenoble Sciences, Les Ulis, EDP Sciences.

Encrenaz, T. & Lequeux, J. (2014) *L'exploration des planètes, de Galilée à nos jours... et au delà*, Paris, Belin.

Brahic, A. & Smith, B. (2015) *Terres d'ailleurs, à la recherche de la vie dans l'Univers*, Odile Jacob

Lequeux, J. & Encrenaz, T. (2015) *À la rencontre des comètes, de Halley à Rosetta*, Paris, Belin: voir le Chapitre 4 pour les matériaux cométaires.

Marty, B. et al. (2016) Origins of volatile elements (H, C, N, noble gases) on Earth and Mars in light of recent results from the ROSETTA cometary mission, *Earth and Planetary Science Letters* 441, 91.

Engrand, C. et al. (2016) Variations in cometary dust composition from *Giotto* to *Rosetta*, clues to their formation mechanisms, *Monthly Notices of the Royal Astronomical Society* 462, 5323.

Fray, N. et al. (2016) High-molecular-weight organic matter in the particles of comet 67P/Churyumov-Gerasimenko, *Nature* 538, 72.

Tartèse, R. et al. (2017) Warm Archean ocean reconstructed from oxygene isotope composition of early-life remnants, *Geochemical Perspective Letters* 3, 55.

Van Kranendonk, M.J., Deamer, D. & Djokic, T. (2017) The new origins of life: Springs' life, *Scientific American* 317, n° 2 (August), 22.

Chapter 11
Exploring Exoplanets: What Prospects?

The discovery of a habitable planet around a star similar to the Sun and the search for signs of life in the atmosphere of this exo-Earth are often presented as the ultimate goal of the quest for exoplanets. Indeed, we can think that the discovery of a planet sheltering a form of life will radically change our point of view on our place in the Universe. But twenty five years of observation of exoplanets, studies in our own Solar System and theoretical work make us be cautious: the discovery of an inhabited exoplanet is probably not for the near future. However, we will see that the way forward is now fairly well signposted, both in terms of the means of observation that should be deployed and in terms of conceptual progress, so that such observations can be envisaged in the decades to come.

Habitable planets, inhabited planets: a difficult problem

One of the latest announcements of the discovery of a potentially habitable planet, that of Proxima Centauri b, illustrates the difficulties of the problem. Proxima Centauri is the closest star from the Sun, at 4.2 light years from us; it is a red dwarf, a M5,5 star, almost ten times less massive and a thousand times less luminous than the Sun, thus very different from it. The planet was discovered following an intense campaign of radial velocity observations, and widely relayed in the social networks under the name of "Pale red dot" (in reference to "Pale blue dot", a famous image taken in 1990 by the Voyager 1 probe, en route to the boundaries of the Solar System: Figure 11.1). With a minimum mass of 1.3 times that of the Earth (remember that the radial velocity method gives only a lower limit of the planet's mass, see chapter 2), it is probably a rocky planet. Its revolution period of 11.2 days corresponds to a distance of 7 million kilometers of the star, which places it in the habitable zone of this very faint star.

FIGURE 11.1 – The "pale blue dot" is this image of the Earth taken in 1990 by the Voyager 1 probe, en route to the boundaries of the Solar System. Our planet occupies only one pixel. The red bands are artifacts of the camera. The book by Carl Sagan *"Pale Blue Dot: A Vision of the Human Future in Space"* (1994) is named after this image. Similarly, the planet discovered in 2016 around the red dwarf Proxima Centauri is called a "pale red dot". (NASA).

Even if this exoplanet is our closest neighbour, we do not know much about it. In particular, as the planet does not seem to transit in front of its star, its radius remains unknown. By making reasonable assumptions about its composition, astronomers have concluded that one could deal either with a large "ocean planet", or with a small rocky planet with a very large metallic core like Mercury in the Solar System (Figure 11.2): two very different kinds of objects! We know even less about the climate of this planet: does it have an atmosphere? Does it always present the same face to its star, hot on this face and icy on the other? One thing only is certain: as Proxima Centauri is a very active star, its planet receives very strong X-ray and ultraviolet radiations, a factor that is probably not very favorable to the appearance of life. Finally, even if the distance from the planet to the star is such that its surface temperature is compatible with the presence of liquid water, it is not impossible that in reality the planet does not contain water at all, especially if it has formed where it is now, so close to the star that all the ice originally present in the protoplanetary nebula was vaporized.

To demonstrate that a "potentially habitable" planet is indeed a "habitable" planet, there are therefore many steps! However, the good news is that in the case of Proxima b, this verification could be possible within a decade. Indeed, the new generation of large ground-based telescopes will be able not only to detect the planet by direct imaging (see Chapter 4) but also to characterize its atmosphere, if there is one. As to whether this planet shelters, or has sheltered, a form of life, this question will be more difficult to solve: let us remember that we still do not know how to answer this question in the case of Mars, which has been studied in great detail by orbiters, probes, and four rovers!

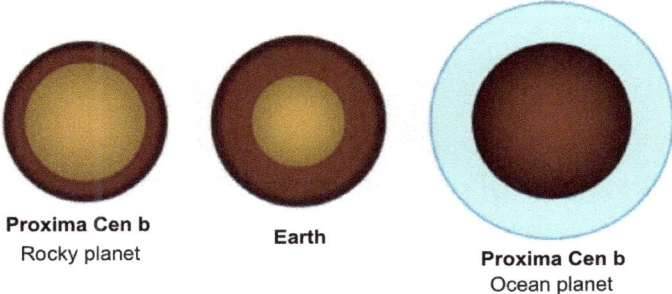

FIGURE 11.2 – Comparison of the two extreme cases proposed for Proxima b with the Earth. This diagram shows the possible internal structure of the planet. From left to right: Proxima b with the smallest possible radius (65% metal core, surrounded by a rocky mantle separated into two phases), Earth (32.5% core), and Proxima b with 50% of rock mantle surrounded by a layer of water in solid and then liquid form. (According to Brugger, B. et al., 2016, *Astrophysical Journal Letters* 831 L16, and the IUU website).

What are the current major issues, in the study of exoplanets that future observation projects will have to address? We know that the Solar System is not unique, since it seems that most stars are surrounded by one or more planets. But it also appears that the Solar System might be singular: for instance, it does not contain any of these super-Earths that seem very common elsewhere, and its planets are almost all in circular orbits. Is it the fact that we have only incomplete samples of exoplanets that suggests that the Solar System is exceptional? We will only know this by observing many more exoplanets, especially small ones or planets located at large distance from their stars, two categories that are very difficult to detect. We also need many more planets with masses and radii precisely measured, the only way to understand their nature and determine their composition. In addition, we are far from understanding the formation and evolution of planetary systems, although it is now clear that migration and instability play an important role. To advance on this point, we need more theoretical studies, but also more observations of young planets, especially giants ones. And finally, we must study the atmospheres of a large number of exoplanets, their composition, their climate, their formation and their evolution: this is a necessary step in the search for signatures of life.

Twenty years ago, the question seemed only technical: it was to build a space telescope capable of measuring biological signatures in the spectrum of an exoplanet atmosphere. We now know that we must first understand the variety of exoplanets to find the best candidates for exo-Earths. Indeed, an inhabited planet will be distinguished only by very subtle clues. It now appears that the quest for exo-Earths is no longer the unique Grail of astronomers: the study of exoplanets resonates with planetology to illuminate the tormented origins of the Solar System and the complex processes that lead to the birth of planets: planetology now extends beyond the Solar System.

More planets, and smaller ones, to explore the diversity of planetary systems

We have learned a lot from the thousands of known exoplanets, but we lack information on certain types. In particular, the census should be extended to smaller objects, for example to determine whether the temperate rocky planets are common or not. The most efficient method of discovery is that of transits: there are more than a dozen programs using ground-based telescopes (Box 11.1), some of which have discovered several dozen planets. In space, after the precursor CoRoT, the NASA mission Kepler, after having discovered several thousand exoplanets candidates, suffered a breakdown of its pointing system. But, after 2014, the satellite has continued its observations in a degraded mode: this is the K2 project, which has provided many additional candidates until the end of the mission in October 2018.

Box 11.1. An example of a transit exoplanet research program: Wide Angle Search for Planets (WASP) and SuperWASP. This project is led by a consortium of 8 universities and institutions in the UK and Spain, which includes two robotic telescopes equipped with wide-angle cameras, one in the Canary Islands (shown here with 8 cameras), the other in South Africa. The program has existed since 2006 and has already discovered more than one hundred exoplanets, confirmed by radial velocities measured in particular with the CORALIE spectrograph of the Haute-Provence Observatory. This is the most productive ground-based program, but it should be noted that in order to achieve this result, it was necessary to observe about 30 million stars! The detection of exoplanets recalls the search for needles in haystacks...

It is essential to confirm by radial velocity measurements the exoplanet candidates discovered by transits from the ground or space. This monitoring is extremely cumbersome and costly in terms of observation time, and above all it is only feasible if the host stars are bright enough, which is not the case for most of the stars discovered by Kepler. NASA's TESS (Transiting Exoplanet Survey Satellite) project, launched on 18 April 2018, targets bright stars to avoid this problem: it will observe 200,000 stars spread over the sky for at least two years, and has already discovered many exo-Earths and super-Earths. Some of these planets should be interesting candidates for transit spectroscopy monitoring, particularly with the future James Webb Space Telescope (JWST).

FIGURE 11.3 – The Transiting Exoplanet Survey Satellite (TESS) launched by NASA in 2018. TESS will observe the entire sky in two years, dividing it into 26 different sectors that will each be observed during 27 days. TESS will observe a sky area 400 times larger than Kepler, and brighter stars; radial velocity monitoring of exoplanet candidates should therefore be less difficult (NASA).

Further in the future (2026), but much more ambitious, the PLATO project of the European Space Agency (ESA) will search for transits of planets on nearly a million stars. Several tens of thousand should be bright enough to allow radial velocity measurements. With its 26 telescopes and exquisite photometric precision, PLATO will have the ability to detect hundreds of rocky planets including analogs of the Earth-Sun system. Moreover, by studying the way in which they vibrate, one can characterize the stars in a very precise way: their mass and age, but also, for the brightest, their radius to better than a few per cent. This information is essential to date the planetary systems and to have good measurements of the radii and masses of the planets. Indeed, the precision on the radius of a planet detected by transit depends directly on the accuracy of the estimate of the stellar radius. This precision is still insufficient to characterize the smallest planets, with a few terrestrial radii. Thus, in general, we cannot say whether they are rocky super-Earths, ocean planets, or even mini-giant planets, mini-Neptunes for example. It is not known, therefore, whether there are really two classes of planets as in the Solar System, rocky and gaseous, or whether there are intermediate types, which would have important consequences for the scenarios of formation of planets.

Another aspect of this problem is that the masses of planets derived from the radial velocity method depend directly on the mass of the star, which is deduced from the spectral type of the star and its luminosity with a large uncertainty. This field should progress enormously with PLATO. With this mission, we should therefore enter the era of precision exo-planetology.

FIGURE 11.4 – The Very Large Telescope (VLT) of the European Southern Observatory. The four 8-m telescopes can be combined either by adding their lights to feed the EXPRESSO instrument, or by interfering with each other and also with four mobile auxiliary telescopes, three of which are shown here, to obtain a high angular resolution and make very precise astrometry. The optical paths are symbolized by white lines superimposed on the photograph. © ESO.

In order to complete the PLATO transit detections, as for all such programs, radial velocities must be obtained with ground-based telescopes: this requires a considerable number of observing nights, an investment that can fortunately be considered on medium-size telescopes. However, monitoring terrestrial candidates around solar-type stars will require the use of large-diameter telescopes such as ESO's Very Large Telescope (VLT) 8-m telescopes (Figure 11.4) or the 10-m GRANTECAN telescope at the Canary Islands, for which specialized high-resolution spectrographs are under development. For example, a consortium of laboratories in Switzerland, Italy, Spain and Portugal has developed the ESPRESSO instrument, which is able to detect velocity variations as small as 10 cm/s on solar-type stars, provided that they are not very active. These performances are very promising if we remember that the presence of the Earth induces a variation of 9 cm/s on the motion of the Sun. ESPRESSO is presently in full operation. To achieve this performance, this instrument combines the light coming from the four 8-meter telescopes of the VLT, making it the virtual equivalent of a 16-m telescope. The main difficulty of the program may not be technical, but may be to find stars that are calm enough so that their activity does not mask the variations in velocity due to the presence of planets. In addition, it will be necessary to convince the committees for allocation of telescope time to give enough time for monitoring exoplanet candidates, as the competition on the large telescopes is always very strong.

As for the future large telescopes such as the European Extremely Large Telescope (E-ELT), a 39-meter giant currently under construction in Chile, the search for exoplanets is one of their priority objectives. Among the instruments that will be mounted on the E-ELT, which should enter in operation around 2025, is a very high-resolution spectrograph even more ambitious than ESPRESSO on the VLT.

FIGURE 11.5 – The European Southern Observatory (ESO) project of a very large telescope, the E-ELT, in construction at the site of Mount Armazones in northern Chile. This new revolutionary telescope concept will have a primary mirror of 39 meters and will be the world's largest optical and near-infrared telescope. It will be a multi-purpose telescope, but its main scientific objectives are exoplanets, the first objects of the Universe, super-massive black holes, as well as the nature and distribution of black matter and black energy that dominate the Universe. Two other extremely large ground-based telescope projects are also currently being developed, the American TMT and the GMT. © ESO.

The space mission CHEOPS of the European Space Agency, which is just entering in full operation, tackles the issue of the masses and radii of exoplanets in an original way. The idea is to observe systems known to shelter planets already detected by velocimetry, and to try to detect their possible transits. The probability is low, but knowing the period of the planets makes it possible to predict the time of the transits and thus to optimize the observation strategy to observe only at that moment.

FIGURE 11.6 – CHEOPS (CHaracterizing Exoplanet Satellite), a project of the European Space Agency, with participations from Switzerland and a dozen other countries in Europe. CHEOPS observes the transit of planets already detected by radial velocities, in the range of mass that goes from super-Earths to Neptune, with the objective of measuring their radii with an accuracy better than 10%. CHEOPS is a small satellite, in the range of 250 kg, and carries a 32-cm telescope. © ESA.

Obtaining masses and radii of the largest possible number of exoplanets is therefore one of the challenges of the next decade, and the object of an impressive astronomical community effort, with dozens of on-going ground-based programs on telescopes of all sizes equipped with state-of-the-art instruments, similar projects on very large ground-based telescopes, and several space missions in operation or development.

Note that in order to enlarge the collections of exoplanets, alongside the transits, another detection technique begins to be fruitful, that of gravitational microlensing (Chapter 5). Unlike the transit method, which favors the detection of planets close to their stars, the planets detected by gravitational lensing can be far from their stars. A dozen microlensing programs are currently under way, with more than 50 detections already achieved. However, as we have seen, information on the host star of these planets is generally completely lacking. The detection of exoplanets by microlensing is one of the objectives of the project of WFIRST (Wide Field InfraRed Telescope survey recently renamed Roman), a NASA space telescope in the near infrared, to be launched around 2025. With its 2.4-m telescope, WFIRST should be able to detect planets as small as Mars, at stellar distances complementary to those detected by transits, and even "floating planets", that is, isolated planets.

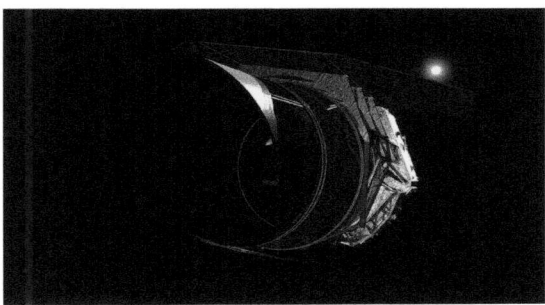

FIGURE 11.7 – NASA's WFIRST wide-field space telescope project, an infrared observatory that focuses on the issue of black energy, the poorly understood phenomenon that accelerates the expansion of the Universe. Another objective is the exploration of exoplanets. WFIRST carries a 2.4-m telescope (the same size as the Hubble Space Telescope), and two instruments, a large-field instrument and a coronograph. The large-field instrument should discover several thousand exoplanets using the gravitational microlensing method, and the coronograph will be used for direct imaging studies of planets. © NASA.

Finally, the radial velocity method has not said its last word for the detection of exoplanets, with instruments like SPIRou (SpectroPolarimeter Infra-Red), which has started operation in February 2019 on the Canada-France-Hawaii telescope. In contrast with existing instruments that operate in visible light, SPIRou performs a large-scale survey of small-mass stars in the infrared, which improves the sensitivity to Earth-like planets by an order of magnitude. SPIRou also measures the mass of new exoplanets that will be discovered by space missions as TESS and PLATO.

New light on giant planets: astrometry and imagery

The first tens of million years of a star are crucial for the formation of its planetary system, and this evolution is largely governed by the giant planets, through their interactions with the protoplanetary disk and migration phenomena (Chapter 6). A large part of the observational efforts of the next decade will therefore concentrate on the long-period giants, which escape detection by transit and are difficult to detect in large numbers by the radial velocity method because of the required observation time. Research will also be devoted to the imaging of the young giants, a growing field thanks to a new generation of ground-based instruments, which should see much progress thanks to two new-generation space telescopes: the JWST from 2021 and WFIRST around 2025.

As for the long-period giants, astronomers expect much from GAIA, the ESA astrometry mission launched in 2013, whose main objective is to trace the history of our Galaxy, the Milky Way, by determining very precisely the position and distance of several billion stars. For this purpose, the relative positions of these stars are measured repeatedly and extremely accurately. If these stars have planets, this translates into periodic variations of these positions. GAIA has an extraordinary astrometric accuracy of about ten microseconds of arc, the thickness of a hair seen at 1000 km. GAIA should thus discover several thousand giant planets, relatively far from their stars (we have seen in Chapter 5 that for astrometry, distant planets are easier to detect), therefore temperate or cold, with periods of revolution that can extend up to the duration of the mission itself, nominally of 5 years but which will certainly be extended. With GAIA, astrometry, the first method used historically to search for planets around other stars, is therefore returning to the forefront. It has already shown statistically that at least 30% of the nearby stars have a massive planet.

FIGURE 11.8 – GAIA, the ESA astrometry mission. GAIA will build the largest 3D-map ever made of our Galaxy, observing repeatedly over a billion stars in the Milky Way, or one percent of all its stars. © ESA.

The direct imaging of exoplanets (see Chapter 4), based on the coronagraphic technique to hide the light from the star, is currently implemented on the ESO Very Large Telescope with the SPHERE instrument, and also with the Gemini Planet Imager (GPI) on the Southern Gemini telescope in

Exploring Exoplanets: What Prospects?

Chile. With these instruments, great progress is being made in the study of young giant planets that are still hot and bright enough in the near infrared to be observed by this technique. The extremely large European telescope, the E-ELT, will also be equipped with imaging capabilities to detect the light emitted by giant planets. In principle, imaging of big Earths or super-Earths should be possible, but it will require substantial progress in instrumentation. Technological research programs are being set up in this direction.

In space, the situation should change by 2021 with the launch and implementation of the JWST (James Webb Space Telescope), developed by NASA in cooperation with the European Space Agency and the Canadian Space Agency. JWST is an infrared observatory with a 6.5-m telescope and 4 instruments. Two of them, NIRCAM and MIRI, supplied by a consortium of European laboratories under the auspices of ESA, are imaging spectrometers with coronographic capabilities, and a third one, NIRISS, is equipped with a device that should also allow to detect exoplanets. These three instruments will thus be able to observe the exoplanets directly, by detecting the light they emit and not by the effects they induce on the motion of their star. All together, they cover a wavelength range that goes from the visible to the mid-infrared. The JWST is also expected to provide information on the poorly known period from the formation of the protoplanetary disk to the appearance of the giant planets because its domain of observation, the mid-infrared range, is particularly well adapted to this question.

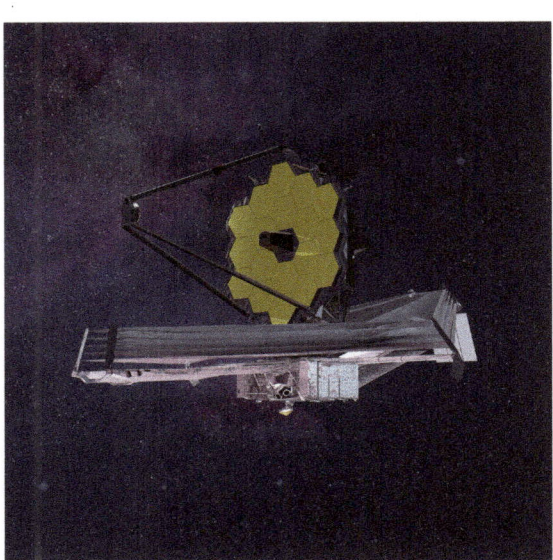

FIGURE 11.9 – The James Webb Space Telescope (JWST), a 6.5-m telescope to be launched in February 2021 by an Ariane 5 rocket from the Guyana Space Center. The JWST is a project of NASA, with a major involvement of ESA and Europe which provide in particular the launch of the spacecraft and two of the instruments. © NASA.

As for WFIRST, which we have already mentioned for its program of detection of exoplanets by microlensing, the addition of a coronagraph is currently under study. In principle, this would allow direct detections of super-Earths to a few astronomical units of their star, and in any case of relatively cold planets, therefore less young than the privileged targets of the JWST.

Studying the atmospheres of exoplanets

Transits, radial velocities, astrometry, microlensing... a whole range of exoplanet detection programs that should greatly increase our understanding of the variety of exoplanets, but also provide a large number of planets with potentially an atmosphere. A great challenge of the next decade is to characterize the atmospheres of these exoplanets: their composition, their structure, the atmospheric circulation, the presence of clouds. Let us take an example in the Solar System: Venus and the Earth. Observed from far, these two planets may seem identical as long as they have not been studied by spectroscopy: then the difference appears between Venus, whose atmosphere reveals only a thick fog of aerosols, and the Earth, with its blue color and the clear signature of the presence of water vapor and oxygen.

If we consider the variety of planets already discovered, it is clear that a good understanding of their atmospheres will not be possible until after a close study of a considerable number of them. The discoveries of the recent years show that the study of atmospheres of hot Jupiters is feasible with the current generation of telescopes on the ground and in space. In the years to come, it should extend to the exoplanets somewhat smaller than Neptune and not too close to their star (temperate mini-Neptunes), as well as to super-Earths. The latter are observable with current instruments, but the uncertainties are great: in general it is not possible to decide whether we are dealing with planets with a rocky core and a thick atmosphere of hydrogen and helium, a micro-Neptune, or an ocean planet with an atmosphere of water vapor... and these worlds are very different. This degeneracy should be more easily solved with some of the observatories and instruments under construction. For exo-Earths and super-Earths, especially if they are temperate, the case becomes more difficult, even with the future very large ground-based telescopes or the JWST.

The characterization of atmospheres can be done in two ways: either, as we have just seen, by direct imaging, rather adapted to giants in distant orbits, or by the transit method, more sensitive to planets close to their stars; these two methods are therefore complementary. For planets in transit, their atmosphere can be studied in different ways: in absorption, by measuring the depth of the transit at different wavelengths, or by following at several wavelengths or by spectroscopy the emission of the planet at different times of its revolution around the star. We have seen in chapters 3 and 8 that these techniques are developing rapidly. In particular, we can detect already the presence of very interesting molecules such as water vapor, methane, carbon monoxide. This is only a very first step in understanding the climate of an exoplanet, for which it is necessary to know, for example, the rotation of the planet, the vertical and horizontal temperature structure of the atmosphere, the wind conditions, the presence of clouds, the presence of oceans, etc.

This type of information can be provided by monitoring the apparent luminosity of the planet throughout its orbit around the star. What we see is the illuminated part of the planet from different angles: the planet presents phases like those of the Moon and Venus. Of course, we cannot image these phases directly, but they translate into a variation in the light we receive (see Figure 3.5). Thus, we can have indications on the presence of an atmosphere (a planet without an atmosphere will have more day/night contrasts in visible light), but also on heterogeneities and clouds, as well as on the redistribution of heat by atmospheric circulation. If there is a redistribution of heat, the temperature of the planet and therefore its emission in the infrared presents less contrast than if there is none. This type of study was carried out with CoRoT and Kepler in visible light, as well as by Spitzer and Hubble in the infrared range. In the future, this technique should provide an idea of the three-dimensional structure of the atmosphere of some planets, which would be a considerable advance.

Observing atmospheres is therefore a very important part of the exoplanet program of the JWST. For this space telescope, it will be easy to follow hot Jupiters, Neptunes and mini-Neptunes, and even hot super-Earths. There is, however, one disadvantage: a complete characterization of the atmosphere will require the use of several instruments in succession, and each for a period longer than that of the transit. These measurements may be very demanding in observation time. Let us take the example of candidates that should be discovered by TESS, which we mentioned earlier. This space mission should provide several candidates for terrestrial planets transiting in front of dwarf stars and located in the habitable zone of those stars less bright than the Sun: they will be temperate terrestrial planets. Such a planet can have a pleasant surface temperature, around 20 °C. If it has an atmosphere, it begins to be very interesting. With a little luck, TESS should detect interesting candidates passing through a dwarf star fairly close to us. These candidates will then have to be confirmed with radial velocity monitoring, before being observed by the JWST. This observation by the JWST should require about a hundred days per planet. So, there is no question of statistics on the nature of the atmospheres of terrestrial planets.

Even if the game is worth it, the JWST is a multi-purpose observatory and it will be necessary to share the time with other important scientific questions like cosmology and the formation of galaxies. If, for example, a quarter of the time of observation is devoted to exoplanets, as is currently the case with Hubble, and over a period of 10 years, a balanced exoplanet program can be envisaged, with the study of the vertical atmospheric structure for a few hundred short-period giants in primary and secondary transit, a more detailed study with a follow-up throughout their orbit of a few dozens of other objects including some super-Earths, a study of the habitability of a handful of temperate terrestrial planets, and the possible search for ozone in some planets sheltering life. A very interesting program, but that does not answer all the questions, far from it!

The need to observe larger samples of planets motivated the scientific community to propose dedicated space missions, with in particular ARIEL (Atmospheric Remote-Sensing Infrared Exoplanet Large survey, Figure 11.10), selected by ESA in 2017 and scheduled for a launch in 2028. It aims to characterize short-period exoplanets around bright stars, including those that will be identified by TESS, CHEOPS and then PLATO. With ARIEL, which will observe in a wide wavelength range from visible to infrared, one could hope, for example, to determine the abundances of carbon and oxygen in the atmospheres of hundreds of giant planets. This would not only be a great step forward in the study of these exoplanets and the understanding of their formation mechanisms, but also to better situate the giant planets of our Solar System in a general context.

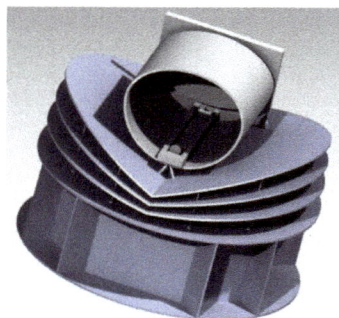

FIGURE 11.10 – A first model of the ARIEL project, designed to study the atmospheres of exoplanets. A 1.1 × 0.7m telescope will be operating from the visible to the mid infrared. © ESA.

The contribution of the future very large ground-based telescopes to transit observations will be more limited because their photometric accuracy cannot compete with that of a space telescope, due to the limitations induced by the Earth's atmosphere. On the other hand, their collecting surface will be much larger than that of the JWST, which has only a 6.50-m mirror; this gives them an enormous advantage for high-resolution spectroscopy. The European E-ELT thus has in its instrumentation plan the development of a spectrograph with very high resolution, HIRES. With such an instrument, we will study the structure in temperature and the dynamics of planet atmospheres. The observation of bio-signatures on rocky planets passing in front of a dwarf M star will in principle be possible in very favorable cases, but probably very difficult both to realize and to interpret, pushing both the instrumentation and our understanding of phenomena at their limits.

Towards analogs of the Earth-Sun system: after 2030?

The projects of observation and characterization of exoplanets are therefore not only numerous, but also very ambitious. It is likely that the next decade will see significant advances in our overall understanding of the field, although they may not be as dramatic as the announcements of the decade following the detection of 51 Pegasus b. One of the final objectives is of course to characterize planets more and more similar to the Earth. But we have seen that the study of temperate terrestrial planets will already be a considerable challenge for the JWST and the very large new-generation ground-based telescopes. With a bit of luck, we will actually study planets of Earth mass orbiting dwarf M stars: these stars are smaller than the Sun, their light is globally redder, and the habitable zone is closer to them. Such planets are at best cousins of the Earth. In fact, the M stars evolve more slowly than the stars of solar type, which can give more time to the life to appear there. But, on the other hand, planets in the habitable zone are probably blocked in a configuration where they always have the same face to their star, which implies large temperature contrasts, hence a rather hostile climate. Solar-type stars (K, G and F dwarfs in order of increasing temperature, in the jargon of astronomers, the Sun being a G dwarf: see Appendix 1) remain therefore privileged targets, and the detection of analogs of the Earth-Sun system is the objective of the PLATO space mission. Given the projected launch date for this mission, starting in 2026, the time required to detect the planets by transit, then to confirm them from the ground by radial velocities, it may be necessary to wait until 2030 to detect a few real exo-Earths, orbiting nearby stars bright enough to consider a detailed study!

Considering these constraints, we may wonder if it is so necessary to look for analogs of the Earth-Sun system. By definition, since the transit period of such systems is of the order of the year, the time required for atmospheric characterization of such objects will take several years or more. It would be better to concentrate on the low-mass stars, around which the habitable exoplanets will have shorter periods. Moreover, there is no reason to believe that the habitable planets are preferentially of the size of the Earth: it is enough that they are rocky, that is to say that they have, a priori, a mass lower than ten terrestrial masses. Temperate super-Earths can therefore be candidates for habitability as well as exo-Earths.

For exo-Earths as for super-Earths, the study of their atmosphere and the search for bio-signatures will require equipments and observatories that we do not yet possess, even if their concepts are discussed and studied since the discovery of the first exoplanets. By definition, exo-Earths and super-Earths are going to be very faint dots, very close to a bright star. We must therefore eliminate as much as possible the light from the star, for example with a coronagraph or with black fringe interferometry, two concepts that have been widely studied with the Darwin projects in Europe and TPF-C and TPF-I (Terrestrial Planet Finder Coronagraph/Interferometer) in the United States (see Chapter 4). Recently, the idea of a new occulting device has emerged. It consists in a space instrument composed of two satellites, one with a telescope and its mirror, the second located at a few tens of thousands of kilometers in order to hide the light from the star (Figure 3.19). Such a concept is currently being studied at NASA under the name of LUVOIR (Large UV/Optical/IR surveyor).

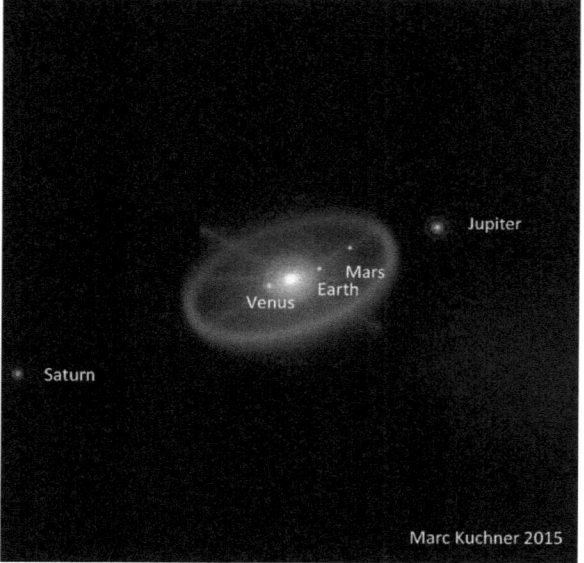

FIGURE 11.11 – Simulated false-color image (450–850 nm) of a planetary system around a star G near like β Cancri, seen by a 12-m space telescope and a coronagraph made of a 100-m screen located at great distance. The imperfections of the screen produce scattering of stellar and solar light around the center of the image, limiting the contrast to $4\ 10^{-11}$ at 1 a.u. Planets like Earth, Venus, Jupiter and Saturn would be well visible. A planet like Mars could even be detected with an integration time of a day as shown here, with a good calibration. We would also see the ring of dust accompanying the asteroids at 3.5 a.u. of the star, ten times fainter than could be detected by the other space missions in progress or planned. After M. Kutner, LUVOIR project.

However, even with the most powerful telescopes, detecting indices of the presence of life will remain very difficult. There is hardly any bio-signature for which we can say with certainty that it is unique, even ozone; in general, there is always an alternative scenario that could explain the observed signals. In order to announce the detection of the presence of life on a planet, we must again have observed a variety of exoplanets, having defined, by comparing the observations with models, what can be considered as a "normal" atmosphere. Then, by contrast, we will know whether the observed anomalies can be interpreted as due to the presence of life. The roadmap to the inhabited exoplanets is thus made up of powerful telescopes and instruments, advances in theory and modeling, but also a lot of patience and some luck!

On the long way to search for life in other solar systems, our own Solar System should provide us with valuable clues. If the detection and study of exoplanets has changed our understanding of the formation and evolution of the Sun and its system of planets, the reverse should be equally true. Let us think of the extraordinary exploration of Comet 67P/Churyumov-Gerasimenko by the Rosetta probe and its lander Philae: the unusual shape and structure of this comet nucleus testify in an eloquent way to the turbulent beginnings of our Solar System. The difficulties we have in interpreting the detection of small quantities of methane in the atmosphere of the planet Mars remind us of the ambiguity of the notion of bio-signature. We will also be able to learn a lot about the frozen worlds by studying the satellites of Jupiter, especially those that possess an ocean under a thick crust of ice like Europe, Callisto and perhaps Ganymede. The study of these icy satellites is the subject of ESA's JUICE (Jupiter ICy moons Explorer) mission and of the Europa Clipper (NASA) mission, both to be launched in 2022 or later.

Alpha Centauri: other planets?

We began this chapter with the discovery, which was widely relayed in the media, of a "potentially habitable" planet around Proxima Centauri, the star closest to us, which is part of a triple system with α Centauri A and B. Are there other planets in this system? The "Breakthrough Starshot" program funded by a Russian billionaire, which aims to send a nanoprobe traveling at 20% of the speed of light to α Centauri, has announced that it will fund ESO for developing one of the instruments of the VLT in Chile with capabilities to image directly the planets around the other two stars of the system. Rendez-vous in 2029 for the first observations... and even if the quest remains unsuccessful with the VLT, these technological advances will be used for METIS, one of the instruments of the E-ELT. If there is a planet the size of Mars in orbit around one of the stars of the system of α Centauri, it will not escape us!

Chapter 12
Communicating with other worlds?

Man has always wondered whether life exists elsewhere, and whether it can be in a sufficiently mature form for communication to be possible. For nearly a hundred years, we have been transmitting radio signals continuously across the Earth's ionosphere, which are betraying a technically advanced civilization. Is it the same on other planets, and can we pick up signals that would come from them, whether emitted involuntarily or intentionally for communicating with us? This question arouses great enthusiasm, and fear also, among scientists and the general public. It is certain that the discovery of extraterrestrial signals would have a great impact on our thoughts and even on our lives. But assuming that advanced civilizations exist on other planets, how can we discover them? And if we find them, what will we do with this discovery?

The probability of existence of advanced civilizations on other planets

This question can be expressed by the famous formula of the American astronomer Frank Drake, posed in 1961, which is to some extent a way of organizing our ignorance. The number N of planets in the Galaxy that harbor a technologically advanced form of life and are capable and willing to communicate is estimated as the product of a number of factors:

$N = n^* \, f_p \, n_h \, f_v \, f_t \, f_c \, (t/10^{10})$, where

n^* is the number of stars in our Galaxy;
f_p is the fraction of stars with planets;
n_h is the average number of habitable planets amongst them;
f_v is the fraction of habitable planets that harbour life;
f_t is the fraction of planets with life where civilisations are technologically advanced;

f_c is the fraction of the preceding planets where civilisations are willing to communicate;

$t/10^{10}$ is the ratio between the average lifetime t in years of the latter civilizations, and the average lifetime of the stars and their planets, estimated roughly at 10^{10} years.

In 1961, the quantities f_p and n_h were totally unknown, as were all of the following factors. Today we have information on f_p (close to 1) and we should be able to specify n_h in the decades to come. We still do not know anything about the following quantities. In his original article, Drake adopted the following values:

$n^* = 10^{11}$ (the whole Galaxy);
$f_p = 0.5$;
$n_h = 2$;
$f_v = 1$;
$f_t = 0.01$;
$f_c = 0.01$;
$t = 10{,}000$ years,

which gives $N = 10$. Given the immensity of the Galaxy, which means that it will probably never be possible to detect signals coming from very large distances, this figure does not leave much hope. But of course many of the values proposed above are eminently debatable, and they have been much discussed! Later, Drake increased very optimistically the probabilities f_t and f_c, which he took equal to 0.3, to arrive at $N \sim t$. However, even so the probability of detection remains very low if the lifetime of communicating civilizations is t $\sim$ 10,000 years, since it is illusory to want to detect too distant signals. If we limit ourselves, for example, to stars nearer than 30 light years, $n^* = 140$ and $n \sim 1.4 \cdot 10^{-9} \, t$, which leaves little hope of detection unless t is really very large. If we consider stars up to 300 light years, we gain a factor of 1000, but the probability is still low.

However, we are in complete unknown with regard to the average lifetime t of an advanced civilization. The example of our own civilization does not encourage optimism: less than two centuries after the first machines appeared, man has built nuclear bombs capable of destroying his civilization, and the overpopulation and global warming of which we are responsible can also destroy it someday. Drake's 10,000 years, if they look short, could be an upper limit of the life span of our civilization. We do not believe Stephen Hawking in considering a migration of man to Mars, where the sojourn would pose insurmountable problems. However, if we are wise enough not to destroy ourselves, we could be victims of the impact of an asteroid or a comet, or of a great volcanic episode, which would eliminate any advanced life on Earth. The last of the five great extinctions of advanced life, that of dinosaurs, dates back 65 million years, and it is estimated that similar phenomena will occur on average every 100 million years. Some attribute these extinctions to volcanism, against which nothing can be done. If, as others think, they are due to the fall of an asteroid, perhaps we will be able to divert this projectile in the future. There is also another limitation: the inescapable increase in the luminosity of the Sun is such that the Earth will leave the habitable zone in about 1.5 billion years, which can be considered as the upper limit of the life of our civilization.

How could we choose between a few hundred or thousands of years, and several hundred million or a billion years for our own civilization? A fortiori, we are totally unaware of what can happen on habitable exoplanets. The only progress since Drake is that we have already found some planets in the habitable zone and that we may have in the near future a more or less complete inventory of these objects near the Sun: so we will soon know where to look. However, it was on the basis of Drake and his colleagues' uncertain conclusions that the first attempts at detection were made, particularly by the SETI (*Search for Extra Terrestrial Intelligence*) projects.

The SETI Projects

In 1959, the physicists Giuseppe Cocconi (1914–2008) and Philip Morrison (1915–2005) studied the best ways to search for signals from possible extraterrestrial civilizations. They concluded that the most promising solution would be to use radio telescopes at the wavelength of 21 cm, that of the line of atomic hydrogen discovered eight years earlier in the interstellar medium. According to them, every advanced civilization would have noticed this property of the most abundant element in the Universe, and would choose this wavelength to communicate.

The following year, Drake and his team launched the OZMA program (named after the princess of the legendary country of Oz), using the 25-meter radio telescope of the National Radio Astronomy Observatory (NRAO) at Green Bank, West Virginia (Figure 12.1). During two months, they observed at 21 cm the stars τ Ceti and ε Eridani, two stars very close, respectively at 3.7 and 3.2 parsecs from us. This was the first such attempt, and the targets were well chosen: we now know that τ Ceti could be surrounded by 5 planets of 2 to 7 terrestrial masses, which still need to be confirmed, some of them in the habitable zone, and that ε Eridani possesses a circumstellar disc and an exoplanet of 3 Jovian masses at 3.4 a.u. from the star, in fact well outside the habitable zone. But no signal was observed.

FIGURE 12.1 – The 25-meter diameter radio telescope of the National Radio Astronomy Observatory (NRAO) at Green Bank (West Virginia). Commissioned in 1959, it is the oldest radio telescope of the NRAO, still equipped with an equatorial mount. © NRAO/AUI/NSF.

After OZMA, other projects were conceived, notably in 1971 the large NASA Cyclops project, which involved the construction of thousands of radio telescopes, with a total surface area of -8 km², that could have detected terrestrial radio and TV signals at distance of 300 pc. Due to its cost, estimated at $ 10 billion, the project has not materialized, but the studies that have been carried out for it are the basis of subsequent achievements. The longest observations made to detect extraterrestrial life were conducted with the Ohio State University radio telescope. On August 15, 1977, a signal was observed with this radio telescope, but it did not reproduce: it was probably a terrestrial emission reflected by some artificial satellite or space debris. Other observations are still being made today at Harvard, the University of California at Berkeley, in Italy and Australia. The last three projects, called *Serendip* (for serendipity), operate in parallel with conventional radio astronomical observations, without requiring extra telescope time. This is a good idea, as radio astronomers (as was one of the authors, J.L.) were very reluctant to devote time from their large radio telescopes to such uncertain searches.

All these projects have been coordinated since 1984 by the SETI Institute, initially financed from 1985 by NASA with $ 1.5 million a year. But NASA stopped funding in 1993 and the Institute is now totally dependent on private funds, with the collaboration of a few American universities. It has piloted and partially funded some fifty different searches for extraterrestrial signals, still without result. Among the most important of these projects is *Phoenix*, which analyzed radio signals between 20 and 10 years ago with various large radio telescopes (Arecibo in Puerto Rico, Green Bank, Jodrell Bank in England, Parkes in Australia and Nançay in France). They looked at a wavelength of 22 cm to 850 solar-type stars located at less than 50 parsecs. In 2007, a network of 42 directional antennas measuring 6.2 m in diameter was commissioned. This is the Allen Telescope Array, initially financed by Paul Allen, one of the co-founders of Microsoft. It is located in Hat Creek, California (Figure 12.2), and could expand to 350 antennas in the future. It observes in priority the stars with exoplanets discovered by the Kepler satellite.

FIGURE 12.2 – Some of the 6.2 m diameter antennas of the Allen Telescope Array. These are eccentric paraboloids, which converge radio waves on a convex secondary mirror which sends them to the focal antenna and receiver, not visible here (Gregory mounting). A metal screen visible under the secondary mirror eliminates the emission from the ground. The covered wavelengths range is from 2 cm to 30 cm. © Shostak S., SETI Institute.

In addition to these searches with radio waves, since 1998 there has been a search program for optical extraterrestrial signals, using a specialized 1.8-meter telescope of Harvard University (the OSETI telescope), built with private funds. This telescope remains fixed overnight, and searches in the sky for very brief optical signals, which would be laser pulses deliberately emitted by extraterrestrial civilizations in the direction of the Solar System. None have been observed so far.

Messages to the Universe

We did not just try to receive messages from exoplanets, but various attempts were made to send messages to them. For example, a message was sent in 1974 by the Arecibo large radio-telescope, 300 m in diameter, in the direction of the globular cluster M 13 which is located at a distance of about 6400 parsecs and contains about 300,000 stars. The power of the emission was several megawatts and the signal could have been easily detected from any point of the Milky Way. The message consisted of 1679 bits distributed in 73 lines of 23 characters each (these are prime numbers, which were supposed to enable the message to be decoded more easily). It thus forms a picture reproduced in Figure 12.3. The problem is that it will take 42,000 years to get a response! Moreover, the deciphering of the message is far from trivial, and it is to be hoped that the extraterrestrials will be more intelligent than us...

FIGURE 12.3 – The Arecibo message. Colors have been added for easier understanding. From top to bottom are shown, in white, from left to right: the numbers 1 to 10 in binary format (the bottom line shows the vertical position of each number). In violet, the numbers 1, 6, 7, 8 and 15 respectively represent hydrogen (H), carbon (C), nitrogen (N), oxygen (O) and phosphorus (P). In green, the empirical formulas of the nucleotides in the configuration incorporated in the DNA. In blue, the double helix of the DNA, the vertical white bar representing the number of nucleotides. In red, silhouette of a man; on the left, in white, the number 14 representing its height (in blue), the wavelength of the message (12.6 cm) serving as a unit. On the right, in white, the human population coded in 32 bits, i.e. 4,292,853,750. In yellow, the Solar System (still with Pluto then considered as a planet), the Earth being slightly offset upwards. In violet, the Arecibo telescope with the trajectoiries of rays, and below its diameter (2430 in wavelength units). The blue lines on each side are apparently meaningless. Wikimedia Commons.

Communicating with other worlds? 175

In the same vein, NASA placed a plate (Figure 12.4) on each of the two Pioneer probes in 1972 and 1973. They may be easier to decipher than the Arecibo message. It reiterated by placing on board the two Voyager probes, which left the Solar System for an infinite voyage in space, a message for possible extraterrestrial civilizations, in the unlikely event that they would recover it. Like the message of Arecibo, it attempts to offer them a description of our own civilization. The Voyager message is in the form of a video record (the Golden Record), which contains a lot of information about the Earth and its inhabitants in the form of images (prudish, we are in the USA!), recordings of noises of animals and cries of infants, and also of different natural sounds. Also included are recordings of greetings in 55 languages, excerpts from literary texts and classical and modern music. Some of the symbols taken from the Pioneer probe plates are engraved on the cover of the disc enclosure, as well as an explanatory diagram of the disc playback mode (Figure 12.5). It is hoped that potential readers will come to understand from these rather obscure schemes how and at what speed they should read the disk, with a stylus included in the probe!

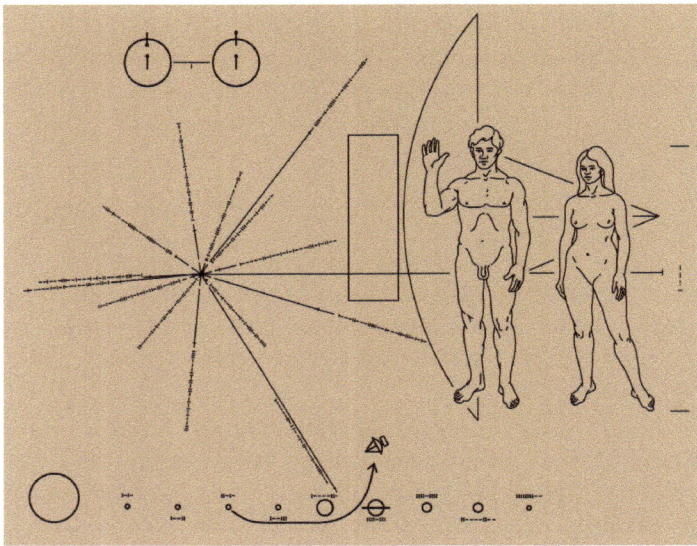

FIGURE 12.4 – The plate placed on board the Pioneer probes, designed by Carl Sagan and Frank Drake. A small diagram in the top left shows that the 21 cm line of hydrogen corresponds to the transition between the two indicated configurations of the H atom, where the spins of the nucleus and the electron are respectively antiparallel and parallel. Below, a diagram showing the direction, distance and period of 14 pulsars. Behind the man and the woman, a diagram of the probe on the same scale. Below, the Solar System, with close to each planet a binary number proportional to its distance to the Sun. We can see the probe leaving the Earth. © NASA.

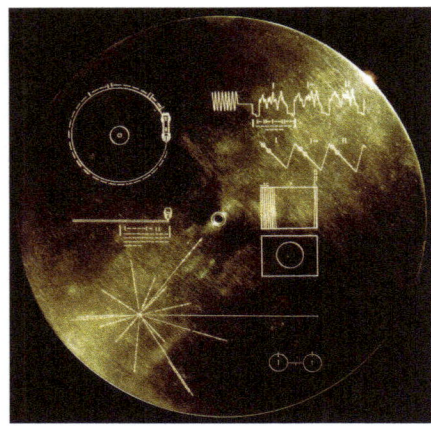

FIGURE 12.5 – The lid of the box that contains the "Golden Record" on the Voyager probes. We find on it some symbols of Figure 12.4, and also a number of diagrams indicating how to read the disc. The story does not say if humans who did not know the technology were asked to understand these patterns. © NASA.

In retrospect, these shipments of bottles to the sea seem very naive. Can we really believe that they will succeed? Their motivation was philosophical, as evidenced by this extract from the official statement of President Carter, placed on the Voyager probes:

> "We cast this message into the cosmos. It is likely to survive a billion years into our future, when our civilization is profoundly altered and the surface of the Earth may be vastly changed. Of the 200 billion stars in the Milky Way galaxy, some – perhaps many – may have inhabited planets and spacefaring civilizations. If one such civilization intercepts Voyager and can understand these recorded contents, here is our message:
>
> This is a present from a small distant world, a token of our sounds, our science, our images, our music, our thoughts, and our feelings. We are attempting to survive our time so we may live into yours. We hope someday, having solved the problems we face, to join a community of galactic civilizations. This record represents our hope and our determination, and our good will in a vast and awesome universe."

While they are probably useless, the messages sent to the Universe did not cost much. However, this is not the case for the attempts to receive extraterrestrial signals. According to estimates by members of the SETI Institute, about $2 million has been invested in the Institute's projects in 2012, and ten times more in all similar activities around the world. Is that reasonable, given that the probability of success is very small? We let the reader make his own opinion. But after all, much more money is being spent on developing more and more sophisticated weapons. The messages involve philosophical, cultural and political considerations, which concern a large public outside the scientific sphere. And yet, we cannot help but dream...

Bibliography

See the website of the SETI Institute: http://www.seti.org/.
Bonnet, R.M. & Woltjer, L. (2008) *Surviving 1000 centuries: can we do it?* Berlin, Springer-Praxis.

Appendix 1:
The planets of the Solar System

Mercury Venus Earth Mars asteroids Jupiter   Saturn Uranus Neptune

Orbital properties

Planet	Semi-major axis (astronomical unit)	Eccentricity	Inclination on the ecliptic (°)	Sideral revolution period (year)	Equatorial radius (km)	Mass (Earth masses)	Sideral rotation period (sideral day)	Inclination of the equator on the ecliptic (°)
Mercury	0.38710	0.20563	7.0048	0.2408	2 439	0.0553	58.646	0.035
Venus	0.72333	0.00677	3.3947	0.6152	6 052	0.8149	243.019	177.36
Earth	1 (1.496 10^6 km)	0.01671	0	1	6 378	1 (5.97 10^{24} kg)	0.99727	23.45
Mars	1.52366	0.09341	1.8506	1.8808	3 396	0.1074	1.0260	25.19
Jupiter	5.20336	0.04839	1.3053	11.862	71 492	317.833	0.4135	3.12
Saturn	9.53707	0.05415	2.4845	29.424	60 268	95.152	0.449	26.73
Uranus	19.2185	0.04638	0.7726	84.020	25 559	14.536	0.71833	97.77
Neptune	30.1104	0.00859	1.7692	164.887	24 764	17.147	0.67125	29.58

Physical properties

Planet	Mass (M_{Earth})	Equatorial radius (R_{Earth})	Density (g/cm^3)	Rotation period	Obliquity of axis (–)	Surface temperature (K)	Atmospheric pressure (bar)	Atmospheric composition
Mercury	0.055	0.382	5.43	58.65 d	0	100/ 700	–	–
Venus	0.815	0.949	5.20	243.0 d	177.33	730	94	96% CO_2, 3% N_2 + CO, SO_2, Ar, H_2O...
Earth	1.000	1.000 6 378 km	5.52	23.93 h	23.45	288	1	78% N_2, 21% O_2 +Ar, CO_2, H_2O...
Mars	0.107	0.532	3.93	24.62 h	25.19	150/ 300	0.006	95% CO_2, 3% N_2 +Ar, O_2, CO, H_2O...
Jupiter	317.9	11.21	1.33	9.92 h	3.08	110 @0.1bar		86% H_2, 13% He +CH_4...
Saturn	95.16	9.45	0.69	10.65 h	26.73	90 @0.1bar		88% H_2, 11% He +CH_4...
Uranus	14.53	4.00	1.32	17.24 h	97.92	55 @0.1bar		84% H_2, 15% He +CH_4...
Neptune	17.14	3.88	1.64	16.11 h	28.80	55 @0.1bar		81% H_2, 18% He +CH_4...

Appendix 2:
A Selection of Exoplanets

The 3600 exoplanets discovered to date present such a diversity of physical and orbital parameters that it is difficult to extract a selection from them. We present below, by way of example, a list of exoplanets and planetary systems having a particular property, either historically or according to one of their parameters. The data are mainly from http://exoplanets.eu, where references to publications are given.

The first exoplanet discovered around a solar-type star

51 Peg b

G2V star, mass 1.11 $M_\odot$, radius 1.266 $R_\odot$, effective temperature 5793 K, distance 14.7 pc.

Planet	Mass	Period	a (semi-major axis of orbit)
51 Peg b	> 0.47 $M_{Jupiter}$	4.2 days	0.052 a.u.

This exoplanet was discovered by velocimetry in 1995 at the Haute-Provence observatory by M. Mayor and D. Queloz. This discovery has challenged our understanding of the formation and evolution of planetary systems.

The first exoplanet discovered by transit

HD 209458 b

G0V star, mass 1.15 $M_\odot$, radius 1.20 $R_\odot$, effective temperature 6092 K, distance 47 pc.

Planet	Mass	Radius	Period	a
HD 209458 b	0.69 $M_{Jupiter}$	1.38 $R_{Jupiter}$	3.52 days	0.047 a.u.

Discovered in 1999, this giant exoplanet is the first to have been detected by primary transit. It has become the prototype of hot Jupiters and has been the subject of multiple measurements of transit spectroscopy. Many atoms and molecules have been detected in its atmosphere.

The first super-Earth discovered very close to its star

CoRoT-7 b

K0V star, mass 0.93 $M_\odot$, radius 0.87 $R_\odot$, effective temperature 5313 K, distance 150 pc.

Planet	Mass	Radius	Period	a
CoRoT-7 b	7.3 M_{Earth}	1.6 R_{Earth}	0.85 days	0.017 u.a.

Discovered in 2009, this planet was for some time the smallest known exoplanet. It probably has the same side facing the star, and the temperature of this face could exceed 1500 K.

The first exoplanet discovered by imaging

AB Pic b

K2V star, mass 0.8 $M_\odot$, radius 0.8 $R_\odot$, effective temperature 4875 K, distance 47.3 pc.

Planet	Mass	Period	a
AB Pic b	13.5 $M_{Jupiter}$	-	275 a.u.

Discovered in 2005 with the NACO instrument of the VLT in Chile by G. Chauvin et al. (*Astronomy & Astrophysics* 438, L29, http://cdsads.u-strasbg.fr/abs/2005A%26A...438L..29C), this exoplanet is the first to have been detected by direct imaging thanks to adaptive optics. Intermediate between a super-Jupiter and a brown dwarf, it probably formed by the direct collapse of a fragment of the primordial interstellar cloud. Spectra were obtained by M. Bonnefoy et al. (2010) *Astronomy & Astrophysics* 512, A52, http://cdsads.u-strasbg.fr/abs/2010A%26A...512A..52B, and give a temperature of about 2000 K.

The nearest exoplanet

Proxima Centauri b

M5.5Ve star, mass 0.12 $M_\odot$, radius 0.141 $R_\odot$, effective temperature 3050 K, luminosity 1.5 10^{-3} $L_\odot$, distance 1.295 pc.

Planet	Mass	Period	a
Proxima Cen b	> 1.3 M_{Earth}	11.186 days	0.0485 a.u.

This planet around the star closest to us was discovered in 2015 by velocimetry with the HARPS and UVES spectrographs of the European Southern Observatory, at the limit of the current instrumental possibilities since the half-amplitude of the radial velocity is only 1.4 m/s. It is in the habitable area and could have an atmosphere, but is subjected to an intense flux of X-rays emitted by the star.

A very dense exoplanet close to its star

Kepler-52 c

M star, mass 0.54 $M_\odot$, radius 0.52 $R_\odot$, effective temperature 4075 K.

Planet	Mass	Radius	Period	a
Kepler-52 c	10.41 $M_{Jupiter}$	0.16 $R_{Jupiter}$	16.39 days	0.103 u.a.

Like his companion Kepler-52 b, this giant exoplanet is close to its star. Discovered in 2012, it has a higher density than iron. It has been suggested that it could be the residue of the high-pressure core of a giant exoplanet formed at a great distance from its star, towards which it has migrated while getting rid of its gaseous envelope. See Mocquet. A. et al. (2014) *Philosophical Transactions of the Royal Society* A, 372, 20130164, http://rsta.royalsocietypublishing.org/content/372/2014/20130164.

A very-low density planet close to its star

XO-6 b

F5 star, mass 1.47 M$_\odot$, radius 1.93 R$_\odot$, effective temperature 6720 K, distance 86 pc.

Planet	Mass	Radius	Period	a
XO-6 b	1.9 M$_{Jupiter}$	2.07 R$_{Jupiter}$	3.76 days	0.0815 a.u.

Discovered in 2016 by transit, this giant exoplanet close to its star has the peculiarity of having a very low density, of the order of 0.2 g/cm^3. The swelling of this type of planet is probably the result of the strong irradiation of which it is the object, which could also lead to an escape of its atmosphere.

A giant exoplanet around a red giant star

L2 Puppis

M5III star, present mass 0.66 M$_\odot$, radius 123 R$_\odot$, effective temperature 3500 K, luminosity 2000 L$_\odot$, [Fe/H] = 0.008, age 10 billion years, distance 64 pc.

Planet	Mass (M$_{Jupiter}$)	a
L2 Pup b (or B?)	12 ± 16	> 2 a.u.

A big planet or perhaps a brown dwarf at a projected distance of 2 a.u. from the star. This star loses a lot of mass forming a very thick circumstellar disc, some of which appears to be being accreted by the planet. This system gives an idea of what the Solar System could be when the Sun will become a red giant, in about 7 billion years: Mercury and Venus will be engulfed by the Sun, while Jupiter could swell by capturing material ejected by the Sun. The fate of the Earth is uncertain. Subsequently, the Sun will become a white dwarf, still surrounded by planets. See Kervella. P. (2016) *Astronomy & Astrophysics* 596, A92.

A compact planetary system around a M star

TRAPPIST-1

M8 star, mass 0.08 M$_\odot$, radius 0.117 R$_\odot$, effective temperature 2560 K, luminosity 5 10^{-4} L$_\odot$, [Fe/H] = 0.04, age 500 million years, distance 12.1 pc.

Appendix 2: A Selection of Exoplanets

Planet	Radius (R_{Earth})	Period (day)	a (a.u.)
TRAPPIST-1 b	1.09	1.510871	0.01111
TRAPPIST-1 c	1.056	2.421823	0.01521
TRAPPIST-1 d	0.772	4.04961	0.02144
TRAPPIST-1 e	0.918	6.09962	0.02817
TRAPPIST-1 f	1.045	9.20669	0.0371
TRAPPIST-1 g	1.128	12.35294	0.0451
TRAPPIST-1 h	0.755	20?	0.63?

Seven Earth-sized planets around a very cold star, all discovered by transit. This is the current record of the number of exoplanets detected and confirmed around a single star. The mass of these planets is determined by their mutual gravitational interactions and is poorly known, but close to that of the Earth as well as their density. Their orbits are practically circular and co-planar. The periods are quasi-resonant, suggesting that these planets were formed far from the star and have migrated. Several of these planets could be in the habitable zone. Absorption spectra of b and c are without characteristic feature, so they are probably purely rocky planets without atmosphere. See Gillon. M. et al. (2017) *Nature* 742, 456.

A planetary system around a G star

Kepler-11 = KOI-157

G star, mass 1.04 $M_\odot$, radius 1.01 $R_\odot$, effective temperature 5836 K, [Fe/H] = 0.06, age 3.5 billion years.

Planet	Mass (M_{Earth})	Radius (R_{Earth})	Period (days)	a (a.u.)
Kepler-11 b	1.9	1.8	10.30375	0.091
Kepler-11 c	2.9	2.87	13.02502	0.106
Kepler-11 d	7.3	3.12	22.68719	0.159
Kepler-11 e	9.5	4.19	31.9959	0.194
Kepler-11 f	2	2.49	46.68876	0.25
Kepler-11 g	300	3.67	118.37774	0.462

Five Earths or super-Earths and a Jupiter discovered by transit around a star analogous to the Sun. Their masses, fairly uncertain, are obtained from the variations of their periods of revolution due to their mutual gravitational interaction. The radii are also somewhat uncertain but these planets all have a low density, weaker than that of the Earth. None is in the habitable zone. See Bedell. M. et al. (2017) *Astrophysical Journal* 839, 94.

4 Jupiters around a hot star

υ Andromedae

F8V star, mass 1.27 $M_\odot$, radius 1.63 $R_\odot$, effective temperature 6212 K, [Fe/H] = 0.09, age 3.8 billion years, distance 13.47 pc.

Planet	Mass ($M_{Jupiter}$)	Period (days)	a (a.u.)	eccentricity e	inclination i
ups And b	> 0.62	4.62	0.059	0.01186	90°?
ups And c	> 1.98	141.26	0.861	0.2445	11.347°
ups And d	> 4.13	1 276.5	2.55	0.316	25.609°
ups And e	> 1.06	3 848.9	5.2456	0.005	?

Four planets around a star already evolved, discovered by velocimetry. They are all Jupiters or super-Jupiters. υ And b is the first exoplanet for which a phase curve has been obtained, showing a temperature difference of 900 K between the illuminated side, which is at 1800 K, and the dark side. Note the large eccentricity of the orbit of c and d and the variety of inclinations of the orbits. The star has a M4.5V companion at 750 a.u., which could be partly responsible for these anomalies.

A system of very massive planets far from their star

HR 8799

A5V star, mass 1.56 $M_\odot$, radius 1.5 $R_\odot$, [Fe/H] = −0.47, age 60 million years, distance 39.4 pc.

Planet	Mass ($M_{Jupiter}$)	Radius ($R_{Jupiter}$)	Period (days)	a (a.u.)	e	i (°)
HR 8799 e	9	—	18,000	14.5	-	—
HR 8799 d	10	1.2	41,054	27	0.1	28
HR 8799 c	10	1.3	82,145	42.9	0	28
HR 8799 b	7	1.2	164,250	68	0	28

Four super-Jupiters on remote orbits discovered by imaging around a fairly young star, in a disk observed with ALMA. They are cooling and are self-luminous. They may have been formed by direct collapse of fragments of the primordial interstellar cloud or outer regions of the disk. Possible resonance b:c:d:e = 8:4:2:1. Animation: https://en.wikipedia.org/wiki/HR_8799#/media/File:HR_8799_Orbiting_Exoplanets.gif

A system of 5 very different planets

55 Cancri

G8V star, mass 0.91 $M_\odot$, radius 0.94 $R_\odot$, effective temperature 5196 K, luminosity 0.58 $L_\odot$, [Fe/H] = 0.31, age 10 billion years, distance 12.34 pc.

Planet	Mass ($M_{Jupiter}$)	Radius ($R_{Jupiter}$)	Period (days)	a (a.u.)	e	i (°)
55 Cnc e	0.026	0.18	0.7365417	0.0156	0.06	85.4
55 Cnc b	> 0.8	—	14.651	0.1134	0.0159	—
55 Cnc c	> 0.169	—	44.3446	0.2403	0.053	—
55 Cnc f	> 0.144	—	260.7	0.781	0.0002	—
55 Cnc d	> 4.802	—	5 218	5.76	0.025	53?

Five very diverse planets, discovered by velocimetry. 55 Cnc e was also observed by transit, and has an atmosphere that could be rich in carbon. The orbits are far from being coplanar. The star around which these planets gravitate, 55 CncA, has a companion located at about 1000 a.u., 55 CncB. See Chapter 5.

A system of six super-Earths or Neptunes

HD 10180

G1V star, mass 1.06 $M_\odot$, rayon 1.05 $R_\odot$, effective temperature 5911 K, luminosity 1.2 $L_\odot$, [Fe/H] = 0.08, age 4.3 billion years, distance 39.0 pc.

Planet	Mass (M_{Earth})	Period (days)	a (a.u.)	e
HD 10180 c	> 13.1	5.75979	0.0641	0.045
HD 10180 d	> 11.75	16.3579	0.1286	0.088

Planet	Mass (M_{Earth})	Period (days)	a (a.u.)	e
HD 10180 e	> 25.1	49.745	0.2699	0.026
HD 10180 f	> 23.9	122.76	0.4929	0.135
HD 10180 g	> 21.4	601.2	1.422	0.19
HD 10180 h	> 64.4	2222	3.4	0.08

Six super-Earths and Neptunes, on rather eccentric orbits.

Appendix 3: Some useful data

Astronomical unit	a.u. = 1.496 10^{11} m
Light-year	l.y. = 9.46 10^{15} m
Parsec	pc = 3.086 10^{16} m = 3.262 l.y.
Mass of the Sun	$M_\odot$ = 1.989 10^{30} kg
Luminosity of the Sun	$L_\odot$ = 3.845 10^{26} W
Effective temperature of the Sun	$T_\odot$ = 5 778 K
Radius of the Sun	$R_\odot$ = 695 700 km
Maas of Jupiter	M_J = 1.899 10^{27} kg
Equatorial radius of Jupiter	R_J = 71 492 km
Density of Jupiter	ρ_J = 1 330 kg m^{-3}
Mass of the Earth	M_T = 5.972 10^{24} kg
Equatorial radius of the Earth	R_T = 6 378 km
Density of the Earth	ρ_T = 5 513 kg m^{-3}
Tropical year	year = 365.242 days = 3.156 10^7 s
Velocity of light	c = 2.997 924 58 10^8 m s^{-1}
Gravitation constant	G = 6.673 10^{-11} N m^2 kg^{-2}
	= 6.673 10^{-8} dyne cm^2 g^{-2}
Stefan-Boltzmann constant	σ = 5.671 10^{-8} W m^{-2} K^{-4}

Data on dwarf stars (solar units)

Spectral type	Radius ($R_\odot$)	Mass ($M_\odot$)	Luminosity ($L_\odot$)	Effective temperature (K)
A5V	2.0	4.0	14.0	8200
F0V	1.4	1.7	6.0	7240
F5V	1.2	1.29	2.5	6540
G0V	1.05	1.10	1.26	6000
G5V	0.93	0.93	0.79	5610
K0V	0.85	0.78	0.40	5150
K5V	0.74	0.69	0.16	4640
M0V	0.63	0.47	$6.3\ 10^{-2}$	3920
M5V	0.32	0.21	$7.9\ 10^{-3}$	3120
M8V	0.13	0.10	$8.0\ 10^{-4}$	2500
Brown dwarf	0.09–0.12	0.013–0.07	$< 2\ 10^{-4}$	< 2000

Appendix 4: The nomenclature of stars and exoplanets

For historical reasons, the nomenclature of stars is extremely complex. A star can have as many as forty different names. To find out, you should consult the free-access database of the SIMBAD data center at Strasbourg (http://simbad.u-strasbg.fr/) where one can find all these different names from any designation, with an enormous amount of data and a bibliography on the star. If we are dealing with a double or multiple star, the names of its different components are identical except that it is completed by the capital letters A, B, and so on.

The nomenclature of exoplanets is much simpler: the planet is designated by the name of its star with the addition of a lower case letter: b, c, d, and so on, according to the order of the discovery, the star itself being implicitly designated by the letter a. The following figure shows what happens in the case of planets gravitating around single or double stars.

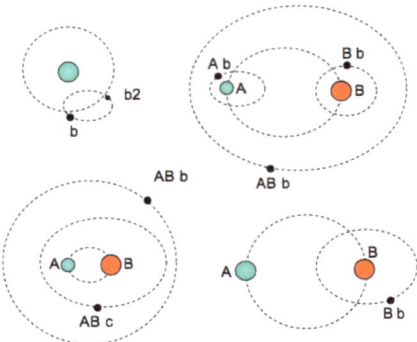

FIGURE A4. – Different possible configurations of orbits of exoplanets, and the suffix of the associated designation. The main star is blue, the possible stellar companion in red and the planet in black. Top left: isolated star with planet b possessing a satellite b2 (no known example, suggested designation). Top right: double star with a planet around each component (e.g. HD 41004) and a planet circling around the whole. Bottom left: two planets around a double star (e.g. NN Ser). Bottom right: a planet around the companion of a double system (e.g. HD 178911). Inspired by Hessman, F.V. et al. (2010).

The International Astronomical Union (IAU) has given names to 31 exoplanets around 14 stars: to see these names, consult http://nameexoworlds.iau.org/names.

Glossary

Accretion: gravitational capture of a solid or gaseous material by a celestial body.

Adaptive optics: technique for correcting images for the effects of atmospheric turbulence.

Albedo: fraction of the light of the central star reflected or diffused by a planet or a comet. One distinguishes the monochromatic albedo, at a given wavelength, and the bolometric albedo or Bond albedo, which concerns the total energy integrated on all wavelengths.

Angular momentum: for a point M, the momentum with respect to a point O of its linear momentum $\boldsymbol{p} = m\boldsymbol{v}$, m being its mass and $\boldsymbol{v}$ its velocity vector, that is to say the vector product $\mathbf{OM} \times \boldsymbol{p}$. For a rotating body, it is the integral of the momenta of its constituent points.

Asteroid: a small solid body belonging to a planetary system; most asteroids of the Solar System gravitate on orbits between Mars and Jupiter.

Astrometry: measurement of the position, the proper motions and the distance of celestial objects (planets, satellites, stars, etc.).

Astronomical unit: unit of length equal to the semi-major axis of the Earth's orbit (around 150 million km).

Brown dwarf: an aborted star, whose mass is insufficient for the nuclear reactions to have been able to start and provide energy. The mass of brown dwarfs is approximately between 0.01 and 0.07 solar mass.

Coma: cloud of gas and dust surrounding the nucleus of a comet.

Comet: an object belonging to a planetary system, on an orbit generally very eccentric and very disturbed by the giant planets. Comets are solid blocks of ice and dark organic matter that evaporate as the Sun approaches, releasing gas and dust responsible for their nebulous appearance and tails.

Continuous spectrum, or *continuum:* emission or absorption covering a wide range of wavelengths without any preferred wavelength.

Coronograph: an instrument that obscures very effectively the light of a star in order to observe weak objects in the vicinity.

Cosmogony: in the general sense, study of the formation and evolution of celestial bodies. This word is usually taken in a restrictive sense and applies only to the Solar System.

Diffraction: a physical optical phenomenon which brings the apparent diameter of the image of a point observed at the wavelength λ to $1.22\lambda/D$ though a circular opening with diameter D.

Doppler-Fizeau (effect): variation $\Delta\lambda$ of the wavelength λ received from a source in motion, with respect to the wavelength λ_0 emitted at rest. We have $\Delta\lambda/\lambda_0 = v/c$, where v is the radial velocity of the source and c the velocity of light.

Eccentricity: for an orbit, deviation from circularity. The eccentricity e of an elliptical orbit is such that the distance from the center to one of the foci of the ellipse is ae, where a is the semi-major axis of the orbit.

Eclipse: see *transit.*

Ecliptic: apparent trajectory of the Sun among the stars during the year.

Equator: for a planet or a satellite, a great circle perpendicular to its axis of rotation, from which the latitude is counted. For the sky, the celestial equator is the projection on the celestial sphere of the equator of the Earth.

Exobiology: study of living matter outside the Earth, or more modestly for the time being of its precursors.

Exoplanet, exocomet: planet or comet gravitating around another star than the Sun.

Gravitational microlensing: a mass like a star or a planet passing in front of a distant star produces a temporary deviation of the light rays by the Einstein effect, which acts as a lens and amplifies the flux of this star. This event is called *gravitational transit*.

Ice line: circle of the circumstellar disk such that water is in the vapor form inside and in the solid form beyond it.

Inclination: angle between the plane of a planetary orbit and a reference plane, which for the Solar System is the plane of the ecliptic. For the orbit of an exoplanet, the plane of reference is the plane of the sky.

Interferometer: An instrument consisting of several interconnected telescopes or radio telescopes (see this word), enabling to achieve a high angular resolution and to map celestial objects.

Interstellar matter: gas and dust filling the Milky Way and other galaxies between the stars. It is the material from which stars and planets form.

Kepler's laws:
1. The orbits of the planets are ellipses in which the star occupies one of the foci.
2. The line segment that joins the planet to the star sweeps equal areas in equal times.
3. The square of the period of revolution of a planet is proportional to the cube of the mean radius of its orbit.

Kuiper belt: region of the Solar System where are located icy celestial objects like Pluto, between 30 and 55 astronomical units of the Sun.

Magnetosphere: the upper region of the atmosphere of a planet with a magnetic field, such as the Earth, Jupiter and Saturn, characterized by the presence of high-energy charged particles trapped in this magnetic field.

Magnitude: a logarithmic scale for measuring the brightness of a star;
- *apparent magnitude:* measures the apparent brightness: $m = -2.5 \log F$ + constant, where F is the luminous flux;
- *absolute magnitude:* measures the intrinsic brightness. By convention, the absolute magnitude M and the apparent magnitude m of a star would be identical if it was at a distance of 10 parsecs (32.6 light years): $m - M = 5 - 5 \log D$, D being the distance of the object in parsecs.

Meridian: for a planet or a satellite, a great circle passing through the poles of rotation. The longitude is the angle between the plane of the meridian of the place and that of a reference meridian, that of Greenwich for the Earth. For the sky, the meridian of a place is the great circle passing through the poles and the zenith (see this word) of this place.

Metallicity: a measure of the abundance of heavy elements in a star, relative to hydrogen. It is generally expressed as the abundance of iron with respect to its solar abundance: $[Fe/H] = \log (Fe/H)_{star} - \log (Fe/H)_{Sun}$

Meteorite: a solid body belonging to the Solar System, whose fall can sometimes be observed on the Earth. The meteorites are very primitive rocks; the carbonacous chondrites are the most interesting meteorites to trace the formation of the Solar System.

Mount:
- *equatorial:* a device for supporting and driving a telescope, one of the axes of which is parallel to the axis of rotation of the Earth; it allows to track a celestial object in its diurnal motion with a single movement.
- *alt-azimuthal:* a similar device but with one axis horizontal and the other vertical. This mount is simpler and less costly than the previous one, but requires a computer control.

Nebula: any astronomical object of diffuse aspect, except comets.
- *Planetary nebulae,* which have long been thought to be at the origin of planetary systems, are actually masses of gas and dust expelled by end-of-life stars.
- *Protosolar nebula:* mass of gas and interstellar dust from which the Solar System was formed.

Nucleosynthesis: formation of chemical elements by nuclear reactions in the stars.

Occultation: see *transit.*

Oort's cloud: area located at the edge of the Solar System, about 1 light year from the Sun, where the comets reside until a gravitational disturbance eventually sends them near the Sun.

Panspermia: the theory that life would have been brought from outside on Earth. In a mitigated form, the theory that only the basic building blocks of life that have been brought.

Parallax: for an object of the Solar System, the parallax is the angle subtended by the equatorial radius of the Earth, seen from this object. It can be measured, for example, by simultaneously photographing the object with respect to the background stars from several points on the Earth. Knowing the radius of the Earth, one can then calculate the distance of the object.

Planetesimal: a small body formed by agglomeration of dust in the protoplanetary disk.

Planetoid: a larger body formed by accretion of planetesimals (see this word) following their mutual collisions; the bigger ones accrete gas and become planets, in the model of Safronov.

Prebiotic molecule: an organic molecule, that is to say formed essentially of the atoms H, C, N and O, which could be a basic brick of the proteins and the DNA. Amino acids are such molecules.

Protoplanetary disk: disk of gas and dust rotating around a proto-star, in which the planets will form. When the planets are formed, they affect the structure of the disk, which is then called *transition disk.* Finally, when all the gas of the disk has disappeared, there remains a *debris disk.*

Pulsar: neutron star in very fast rotation, regularly emitting, like a beacon, a radio signal whose period is extremely stable.

Radial velocity: the velocity of approach or recession of an object, measured by the Doppler-Fizeau effect on its spectral lines.

Radio astronomy: branch of astronomy which consists of studying the radio emission of celestial objects. The Sun, planets, certain stars, the atomic,

Glossary 197

molecular or ionized interstellar gas, high-energy electrons of the cosmic radiation, pulsars, galaxies and quasars emit radio waves.

Radio telescope: antenna or set of antennas used for radio astronomy.

Refractor, or *refracting telescope:* an instrument of observation whose objective is a single or double lens, which makes the image of the sky at its focus where it is observed with an eyepiece.

Revolution: movement of a planet around its star, or a satellite around its planet.

Rotation: movement of a planet or a satellite around its axis.

Solar wind: ionized gas continuously ejected by the Sun's corona.

Spectral band: a set of often unresolved spectral lines that produce absorption or emission covering an extended range of wavelengths. The bands are characteristic of the molecules, which produce very many lines.

Spectral line: increase or decrease of intensity in the spectrum of an object occurring at a given wavelength; the line is called emission if there is reinforcement, and absorption if there is decrease. The wavelength of a line is characteristic of the atom, ion or molecule which produces it.

Spectroscopy: technique consisting in decomposing by a prism or a diffraction grating the light in its different wavelengths. By extension, decomposition of any electromagnetic wave (ultraviolet, infrared or radio). The instrument used is called *spectroscope* if it is not recording, or *spectrograph* if it is. Spectroscopy can be *in emission* if one observes a luminous body as an exoplanet, or *in transmission* if one observes for example the atmosphere of an exoplanet interposed in front of a star.

Supernova: massive star ending its life in an explosion.

Telescope: an observation instrument whose objective makes the image of the sky at its focus, where it is observed with an eyepiece or another device. The word was used initially to designate an instrument with an objective made of one or two lenses, but it is now often used also for an instrument whose objective is a mirror. The *Schmidt telescope* is an instrument with a spherical mirror equipped with a plate to correct the spherical aberration, which gives it a large field of view.

Terrestrial planet: a rocky planet of the Solar System comparable to the Earth: Mercury, Venus, Earth and Mars, as opposed to the giant planets (Jupiter, Saturn, Uranus and Neptune). This term can be extended to exoplanets.

Tide: deformation or rupture of an object under the effect of the gravitation of a neighboring star.

Transit: passage of a planet in front of its star (*primary transit*) producing an eclipse of the star by the planet, or passage behind the star (*secondary transit*), producing an occultation of the planet by the star. *Gravitational transit*: see gravitational microlensing.

Velocimetry: method of indirect detection of an exoplanet by the variations of radial velocity that it produces on its star.

Acronyms

ARIEL: Atmospheric Remote-sensing Infrared Large survey (ESA satellite)

A-STEP: Antarctic Search for Transiting ExoPlanets

AST3: Antarctic Survey Telescopes

CFHT: Canada-France-Hawaii Telescope

CHEOPS: CHaracterizing ExOPlanet Satellite (European)

CORALIE: Spectrometer for radial velocities of the Swiss telescope at ESO; this is not an acronym, but the name derives from its predecessor CORAVEL (CORrelation of VELocities)

CoRoT: COnvection, ROtation et Transits planétaires (French satellite)

CSTAR: Chinese Small Telescope ARray

ELT: Extremely Large Telescope

E-ELT: European Extremely Large Telescope (of ESO)

ESA: European Space Agency

ESO: European Southern Observatory

ESPRESSO: Echelle SPectrograph for Rocky Exoplanet and Stable Spectroscopic Observations (at ESO)

GAIA: Global Astrometric Interferometer for Astrophysics (ESA; the method is no longer interferometric, but the acronym has been retained)

GMT: Giant Magellan Telescope (American)

GPI: Gemini Planet Imageur (American)

GRANTECAN: GRAN TElescopio CANarias (Spanish telescope with a diameter of 10.4 m)

HARPS: High Accuracy Radial velocity Planet Searcher (spectrograph at ESO)

HAT: Hungarian Automatic Telescope

HIRES: High REsolution Spectrometer (on the E-ELT)

HST: Hubble Space Telescope

JUICE: JUpiter ICy moons Explorer (ESA project)

JWST: James Webb Space Telescope

K2: Kepler 2, second observation program with this satellite

LUVOIR: Large UV/Optical/IR surveyor (American project)

MASCARA: Multi-site All-Sky CAmeRA (ESO-Leiden search for exoplanets)

METIS: Mid-infrared E-ELT Imager and Spectrograph

MIRI: Mid-InfraRed Instrument (on the JWST)

MMT: Multi-Mirror Telescope (American)

NACO: contraction of NAos-COnica, an ESO adaptive optics instrument resulting from the combination of NAOS (Nasmyth Adaptive Optics System) and CONICA: COudé Near-Infrared CAmera (ESO)

NASA: NAtional Space Administration (USA)

NIRCAM: Near-InfraRed Camera (on the JWST)

NIRISS: Near-InfraRed Imager and Slitless Spectrograph (on the JWST)

NRAO: National Radio Astronomy Observatory (American)

OGLE: Optical Gravitational Lensing Experiment

OSETI: Optical Search for Extra Terrestrial Intelligence (American telescope)

PLATO: PLAnetary Transits and Oscillations of stars (ESA project)

SETI: Search for Extra Terrestrial Intelligence (refers to all such searches)

SOPHIE: Spectrographe pour l'Observation des Phénomènes des Intérieurs stellaires et des Exoplanètes (at the Haute-Provence observatory)

SPHERE: Spectro-Polarimetric High-contrast Exoplanet REsearch (ESO instrument)

SPIRou: Spectro Polarimètre InfraRouge (ar the CFHT)

TESS: Transiting Exoplanet Survey Satellite (American project)

TMT: Thirty-Meter Telescope (American)

TOPS: Towards Other Planetary Systems (American report)

TPF-C: Terrestrial Planet Finder Coronograph (abandoned American project)

TPF-I: Terrestrial Planet Finder Interferometer (abandoned American project)

TRAPPIST: TRAnsiting Planets and PlanetesImals Small Telescope (2 Belgian telescopes, at Chili and Morocco)

TTV: Transit Timing Variation (exoplanet detection method)
UVES: UltraViolet Echelle Spectrograph (at the ESO VLT)
VLT: Very Large Telescope (4 telescopes 8 m in diameter at ESO)
WASP: Wide Angle Search for Planets
WFIRST: Wide Field Infrared Survey Telescope (American)

Index

A

Adaptive optics 49, 53-56, 78, 124, 182, 193, 200
  NACO 54-55, 182, 200
Amino acid 131, 132, 135, 136, 138-140, 148, 196
Albedo 72, 109, 193
Asteroid 9, 11, 92, 96, 98, 100, 123, 124, 133, 135, 166, 170, 179, 193
Astrometry (detection by) 15, 21, 23, 40, 49, 50, 56, 156, 159, 160, 162, 193
Atmosphere of exoplanets 2, 40-42, 44, 56, 57, 64, 72, 73, 107-119, 131, 133-135, 140, 145, 145-149, 151-154, 162-167, 182-185, 187, 195, 197
Atmospheric motions 115, 118, 120, 121

B

Babinet, Jacques (1794-1872) 9
Bacteria 132-135, 140, 142-144, 146
Bada, Geoffrey 148
Barnard, Edward E. (1857-1923) 14
Bergerac, Cyrano de (1619-1655) 6
Bion, Nicolas (1652-1733) 7
Brown dwarf (star) 16, 38, 44, 45, 55, 62, 64, 67, 77, 100, 116, 119, 128, 182, 184, 190, 193
Bruno, Giordano (1548-1600) 1, 5-8
Buffon (1707-1788) 8, 9
Butler, Paul 27, 30, 31

C

Cameron, Alastair G.W. (1925-2005) 11, 78
Chamberlain, Thomas (1843-1928) 10

Chaos 2, 101, 102, 105
Charbonneau, David 34, 35, 46
Chirality 139, 140
Chlorophyll 140, 144, 146, 147
Classification of exoplanets 73, 109, 112, 191
Cocconi, Giuseppe (1914-2008) 171
Comet 9, 12, 92, 95, 96, 98, 110, 123-127, 130, 132-136, 167, 170, 193, 194, 196
Coronograph 50, 51, 55, 56, 124, 159, 161, 193
Cuze ou Cues, Nicolas de 1, 5
Cyclops (project) 172

D

Descartes, René (1596-1650) 6-10
Direct detection: see **Imaging of exoplanets**
Disk: protoplanetary, transition, debris 2, 9, 10-12, 19, 20, 26, 59, 66, 75-87, 89-104, 123, 124-127, 133, 159, 161, 187, 194
DNA 133, 136, 137-139, 148, 174, 196
Doppler-Fizeau effect 20-22, 118, 125, 194, 196
Drake, Frank D. 8, 169-171, 175
Dust trap 90, 91

E

Earths and habitable planets 70-73
Eclipse: see transit, secondary
Epicurus (ca. 342-279 BC) 5, 6
Evolution of Solar System 92-102
Eccentricity 26, 27, 60, 61, 100-103, 179, 186, 194
Exocomet 124-127, 194
Exobiology 131-149, 194

F

Flammarion, Camille (1842-1925) 8
Fontenelle, Bernard le Bovier de (1657-1757) 6-8
Formation of Solar System 8-10
Formation of stars and disks 75-87
Formation and evolution of planetary systems 89-105
Fraunhofer, Joseph von (1787-1830) 13

G

Goldwin, Francis (1562-1633) 6
Great late bombardment 95, 97, 98, 103, 133, 135, 142

H

Habitability and life in the Solar System 6-8, 70, 140-144, 157
Habitability and life on exoplanets 38, 39, 70-72, 103, 110, 144-149, 151-152, 164-165, 169-171, 183, 185
Haldane, John (1892-1964) 131
Hawking, Stephen 170
Henry, Gregory 34
Herschel, William (1738-1822) 8-10, 13, 14
Holmberg, Erik (1908-2000) 14
Huygens, Christiaan (1629-1695) 8

I

Ice line 78, 82, 83, 89, 92, 93, 103, 104, 110, 193
Imaging of exoplanets 50, 55-57, 119, 122, 154, 159-162, 182
Imaging systems for exoplanets
  GPI 160, 199
  METIS 167, 200
  NIRCAM 161, 200
  NIRISS 161, 200
  SPHERE 56, 81, 160, 200
Interaction planet-disk 84-87
Interferometer 50, 52, 53, 78, 194

J

Jacob, capitaine William S. (1813-1862) 14
Jet (protostellar) 75-78
Jupiters hot and cold 64-67

K

Kamp, Peter van de (1901-1995) 14
Kant, Emmanuel (1724-1804) 9, 10, 75, 78
Kasting, James 70
Kepler, Johannes (1571-1630) 6, 8, 26
Kepler's laws 26, 47, 194
Kuiper's belt 95-100, 123, 126, 127, 195

L

Laplace, Pierre-Simon (1749-1613) 2, 9-10, 75, 78, 100
Lipid 138, 139, 148
Lucretius (ca. 94-54 BC) 5
Lyot, Bernard (1897-1952) 51, 52

M

Magnetosphere 57, 195
Marcy, Geoffrey 27
Marois, Christian 119
Mayor, Michel 24-26, 67-69, 181
Miller, Stanley (1930-2007) 131, 135, 139, 148, 149
Messages to Universe 174-177
Metallicity 65, 68, 201
Meteorite 91, 92, 132-136, 138-140, 142, 195
Migration 2, 60, 84-87, 89, 90, 93-95, 97, 103, 104, 153, 159
Molecule: cometary 132, 135
  : interstellar 76, 78, 131
  : pre-biotic 130-140, 148-149, 196
Morrison, Philip (1915-2005) 171
Moulton, Forest Ray (1872-1952) 10, 14

N

Nebula: Orion 76
  : planetary 9, 10, 195
Newton, Isaac (1642-1727) 6, 8
Nice model 92-98
Ninth planet 98, 99
Nomenclature of stars and exoplanets 191, 192
Nucleotide 136, 137, 148, 174

O

Occultation: see transit, primary
Oort's cloud: 12, 95, 196

Index

Oparine, Alexandre (1894-1980) 131
Oró, Juan (1923-2004) 148
OZMA (project) 171, 172
Ozone 145-147, 164, 167

P

Panspermia 132-133, 196
Peptide 135, 136, 140
Phoenix (project) 172
Piazzi, Giuseppe (1746-1826) 9
Planetesimal 11, 78, 79, 89-91, 98, 104, 196
Planetoid 11, 78, 79, 89-93, 133, 134, 196
Protein 133-136, 138-140, 196
Pulsar 1, 19-21, 175, 196
  PSR B1257+12 19-21
  PSR B1620-26 62

Q

Queloz, Didier 24-26, 181

R

Radioastronomy (detection by) 57
Radiotelescope 12, 171-174, 194, 197
  Allen Telescope Array 173
  ALMA 79-83, 123, 127, 187
  Arecibo 19, 173-175
  NRAO 25-m 171, 172
  Ohio State University 172
  SKA 57
  UTR-2 (Kharkiv) 57
Resonance 87, 94-98, 100, 101, 103, 104, 187
Reuyl, Dirk (1906-1972) 14
Ring around exoplanet 127, 128
Rotation of exoplanets 72, 109, 121, 197

S

Safronov, Viktor (1917-1999) 11, 89, 196
Sagan, Carl (1934-1996) 152, 175
Satellites: artificial satellites and space probes
  ARIEL 44, 164, 199
  Cassini 100, 164
  CHEOPS 44, 158, 164, 199
  CoRoT 28, 33, 37-39, 154, 199
  DARWIN 52, 166
  Euclid 46
  Europa Clipper 167
  GAIA 40, 49, 50, 160, 199
  Galileo 146
  Herschel 83, 126, 127
  Hubble Space Telescope (HST) 12, 42-45, 77, 78, 111, 112, 123, 163, 200
  IRAS 12, 19
  ISO 22
  JUICE 167, 200
  JWST 44, 121, 147, 155, 161-165, 200
  Kepler 28, 33, 36, 38, 39, 44, 48, 59, 64, 129, 154, 155, 163, 173, 200
  LUVOIR 166, 200
  Odin 126
  Pioneer 175
  PLATO 155, 156, 159, 164, 165, 200
  Spitzer 12, 42, 43, 104, 111, 120, 121, 126, 163
  TESS 44, 155, 159, 163, 164, 200
  TPF 166, 200
  Voyager 131, 143, 151, 152, 175, 176
  WFIRST 46, 159, 160, 162, 201
See, Thomas J.J. (1866-1962) 14
Serendip (projects) 172
SETI 144, 171-173, 176
Snellen, I. 118
Spitzer, L. (1914-1997) 10
Strand, Kaj Aa. (1907-2000) 14, 15
Struve, Otto (1897-1963) 15, 16, 24
Spectrometer for exoplanets 23, 29, 197
  CORALIE 67, 154, 199
  CORAVEL 199
  CRIRES 118
  ELODIE 25
  ESPRESSO 29, 156, 157, 199
  HARPS 28, 67, 183, 199
  HIRES 28, 164, 200
  METIS 174, 206
  MIRI 147, 161, 200
  NIRCAM 161, 200
  SOPHIE 64, 200
  SPHERE 50, 56, 81, 160, 200
  SPIRou 30, 159, 200
  UVES 183, 201

Spectroscopy: emission 43, 44, 111,
    112, 155, 162, 164, 182, 197
    : transmission 41, 42, 64,
    111, 112, 124, 155, 164, 182, 197
Spitzer, Lyman (1914-1997) 7
Stars (data on) 190
Stars, individual
    2MASS J14074792-3945427 128
    47 Ursae Majoris 27
    51 Pegasi 1, 24, 25, 27, 59, 60, 118,
        165, 181
    55 Cancri 61, 62, 187,
    61 Cygni 14
    70 Ophiuchi 13-14
    70 Virginis 27
    α Centauri 167
    α Lyrae (Véga) 12
    α Piscis Austrinus (Fomalhaut) 123
    β Pictoris 12, 55, 57, 124-127
    ε Eridani 171
    ζ Aquarii 14
    η Corvi 126
    μ Draconis 14
    ξ Bootis 14
    τ Bootis 121
    τ Ceti 171
    υ Andromedae 186
    AB Pic 55, 182, 183
    Barnard (étoile de) 14
    CoRoT-7 38, 108, 182
    CoRoT-9 38
    CoRoT-15 38
    GJ 1214 116, 117
    GJ 3470 118
    GL 436 116
    HD 3651 27
    HD 10180 187, 188
    HD 97048 81
    HD 114762 16
    HD 131399 56
    HD 163296 82
    HD 172755 126
    HD 176051 15
    HD 189733 42, 43, 111-115, 120
    HD 209458 (Osiris) 34, 35, 42, 109,
        111-115, 118, 181, 182
    HH 30 77
    HL Tauri 12, 79-81, 87
    HR 8799 57, 87, 119, 187, 188

    Kepler-11 = KOI-157 185
    Kepler-20 38
    Kepler-52 108, 183
    Kepler-57 108
    L2 Puppis 184
    Proxima Centauri 14, 151-153, 167,
        183
    RX J1615 81
    SAO 206462 81
    TRAPPIST-1 35, 87, 147, 184, 185
    TW Hydrae 80, 81, 87
    V883 Orionis 83
    XO-6 184
Super-Earths and Neptunes 67-70, 98
Survey for exoplanets
    A-STEP 36, 199
    AST3 36, 199
    CSTAR 36, 199
    HAT/HATnet 36, 199
    OGLE 36, 45, 200
    OSETI 173, 200
    TRAPPIST 201
    WASP/SuperWASP 36, 154, 201

T
Telescope (reflecting) 12, 23, 152, 194,
    195
    Canada-France-Hawaii (CFH) 30,
        100, 159, 199
    ESO 3,6m 28, 67, 124
    ESO E-ELT 29, 157, 161, 199
    ESO VLT 29, 35, 55, 56, 76, 118,
        119, 124, 156, 157, 160, 201
    Gemini 56, 119, 160
    Giant Magellan Telescope (GMT)
        29, 157, 199
    GRANTECAN 156, 199
    Keck 28, 119
    MMT (Multi-Mirror Telescope)
        36, 200
    Observatoire de Haute Provence
        25, 64
    Subaru 56
    Swiss at ESO 67, 199
    Thirty Meter Telescope (TMT) 29,
        157, 200
Telescope (refracting) 13, 196, 197, 202
    Dorpat (Tartu) 13
    Sproul 14

Index

Temperature of exoplanets 70-73, 109-116, 118-121, 163-165
TOPS (report) 17, 49, 51, 200
Transit: detection by 33-40, 47, 197
: gravitational (detection by) 44-46, 48, 194
: primary 41-42, 197
: secondary 43, 197
: **Transit Timing Variation (TTV): detection by** 40, 63, 201

U

Urey, Harold (1893-1981) 131, 135, 139, 148, 140

V

Velocimetry: detection by 19-31, 49, 204
Voltaire (1694-1778) 6
Vortex model of planetary systems 7, 10

W

Weiszäcker, Carl Friedrich von (1912-2007) 10
Whipple, Fred (1906-2004) 90
Wilkins, John (1614-1672) 6
Wolszczan, Alexander 19, 20

www.ingramcontent.com/pod-product-compliance
Ingram Content Group UK Ltd.
Pitfield, Milton Keynes, MK11 3LW, UK
UKHW061222180426
11947UKWH00026B/1977